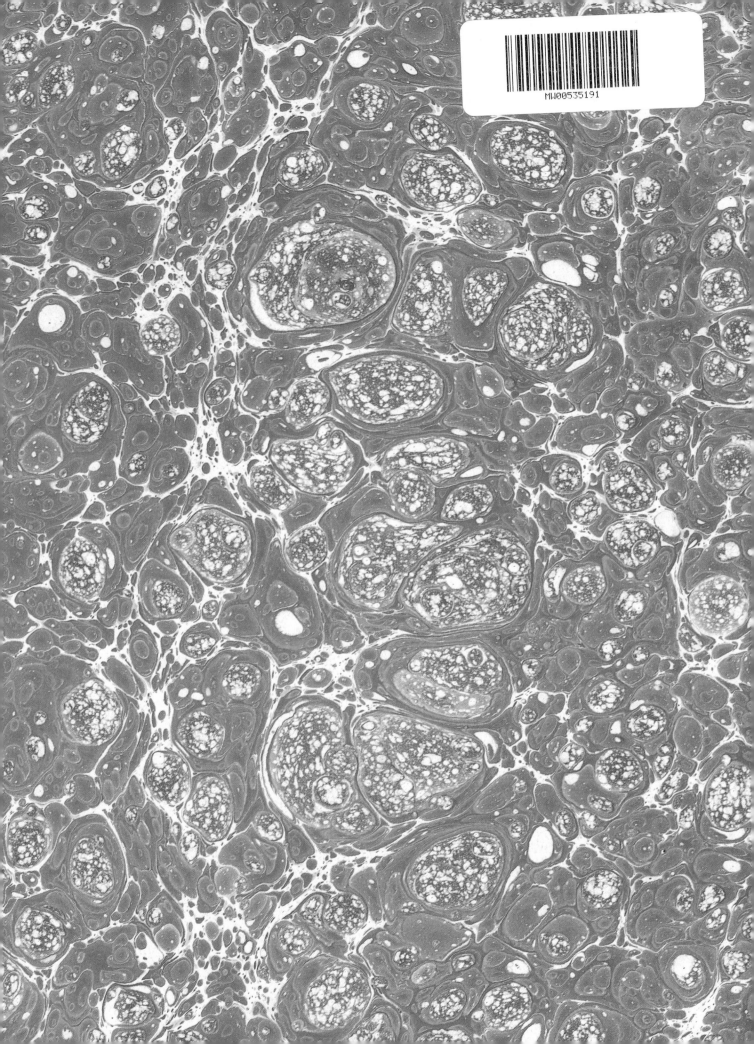

Werner Müller

THE HEAVY FLAK GUNS
1933-1945

THE HEAVY FLAK GUNS

1933-1945

• 88mm • 105mm • 128mm • 150mm •

AND BALLISTIC DIRECTIONAL EQUIPMENT

Werner Müller

1469 Morstein Road, West Chester, Pennsylvania 19380

BIBLIOGRAPHY

Engelmann-Scheibert, *Deutsche Artillerie 1934-bis 1945*,
C.A. Starke-Verlag

Hogg, J., *Luftwaffendienstvorschriften*,
Motorbuch-Verlag

Koch, H.A., *FLAK*, Podzun-Verlag

v. Renz, O.W., *Deutsche Flugabwehr im 20. Jahrhundert*,
Mitter-Verlag

Senger-Etterlin, *Die Deutschen Geschütze 1939 bis 1945*,
Lehmans-Verlag

PHOTO CREDITS

Translated from the German by Dr. Edward Force,
Central Connecticut State University.

Copyright © 1990 by Schiffer Publishing.
Library of Congress Catalog Number: 90-61169.

Printed in the United States of America.
ISBN: 0-88740-263-1

This book originally published under the title,
Die Schwere Flak 1933-1945,
by Podzun-Pallas Verlag, GmbH, Markt 9, 6360 Friedberg 3
© 1990. ISBN: 3-7909-0230-6.

We are interested in hearing from authors with
book ideas on German military history.

Published by Schiffer Publishing, Ltd.
1469 Morstein Road
West Chester, Pennsylvania 19380
Please write for a free catalog.
This book may be purchased from the publisher.
Please include $2.00 postage.
Try your bookstore first.

CONTENTS

P.7002g

After 1918 the "train battalions"—that was the name that concealed the anti-aircraft units, which were banned at that time—were armed with the 75mm L/60 anti-aircraft gun (above). These guns, though, were withdrawn from service and sold abroad after the development of the 88mm Flak 18.

88mm FLAK 18, 36, 37

The use of warplanes in warfare necessitated the development of weapons to be used against them. Up to the end of World War II, the following heavy anti-aircraft guns were used: the 88mm, 105mm and 128mm Flak. A 150mm gun existed only as a prototype and was never put into series production.

After World War I the German Army was banned from using any anti-aircraft weapons and accompanying fire-control devices developed up to that time. After the Geneva Convention of 1932, at which the German Reich was recognized as having an equal standing to other sovereign states, the development of an anti-aircraft troop within the army could begin. As early as 1933 the 75mm Flak L/60 guns used until then were gradually replaced by the 88mm Flak 18. The model designation "18" was applied to all these weapons to conceal the developments that had been carried on after the arms limitation of 1920.

Until the introduction of universal military service in 1935, the formation and development went on within the army under the deceptive term of "Fahrabteilung"—the British counterpart being "train battalions." Only at this point were the Flak units removed from the Army and transferred to the newly-formed Luftwaffe. In November of 1935 these units were given the blue-gray Luftwaffe uniform with the red service-arm color of the Artillery on the collar patches.

Of all Flak weapons used in World War II, the 88mm Flak was generally the weapon most feared by the enemy on land and in the air, and most desired by the German troops to support them in ground warfare, repel tanks and attack bunkers.

The pivoting mount of the 88mm Flak with its octagonal platform rested on four arms. This pillar-type mount allowed a vertical field from -5 to +85 degrees and an unlimited horizontal field. The barrel consisted of a jacket with unscrewable breech and full-length one-piece inner barrel. The independent, self-setting shearing-handle lock opened when the barrel was advanced. Thus the shell was ejected and the spring compressed again.

The loading tray with the rammer did not work well and was usually removed by the troops. The barrel, which recoiled when fired, was halted by the recoil brake, which lay in the cradle under the barrel, and moved back to its original position by the pneumatic recuperator. (See drawing showing functioning of recuperator). In two other cylinders under the barrel were the spring equalizers, which made the vertical adjustment of the gun easier.

88mm Flak 18 ready for transportation.

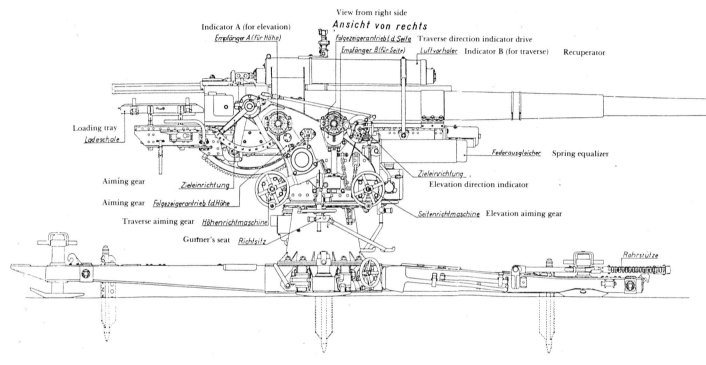

View from right side
Ansicht von rechts

Indicator A (for elevation)
Empfänger A (für Höhe)

Folgezeigerantrieb f.d Seite Traverse direction indicator drive

Empfänger B (für Seite) *Luftvorholer* Indicator B (for traverse) Recuperator

Loading tray
Ladeschale

Federausgleicher Spring equalizer

Aiming gear

Zieleinrichtung *Zieleinrichtung* Elevation direction indicator

Aiming gear *Folgezeigerantrieb f.d Höhe*

Seitenrichtmaschine Elevation aiming gear

Traverse aiming gear *Höhenrichtmaschine*

Gunner's seat *Richtsitz*

Rohrstütze

Three views of the 88mm Flak 18 from the right side. On this side of the gun were the elevation and traverse aiming gear, with the targeting mechanism for land targets and the lamp or pointer direction indicator receiver. These indicated the firing data given by the fire control device.

The elevation, traverse and fuze settings were provided by the fire control device via a 108-strand remote-control cable to the lamp transmission device 30. These settings could also be transmitted by telephone. The fuze setter, at first separated from the gun, was later mounted directly on the gun mount. The gun was portable through the use of a two-axle special trailer 201, its rear axle carrying dual wheels, which meant that the two ends were not interchangeable. When on this limber, the gun pointed in the direction it was moving.

The drawing in the middle shows the left side of an 88mm Flak 18. In the longitudinal view below, note particularly the horizontal adjustment mechanism in the mount. With it the gun could be turned up to five degrees in either direction vertically. Information on the operation of the pneumatic recuperator, recoil brake and spring equalizer will be provided later.

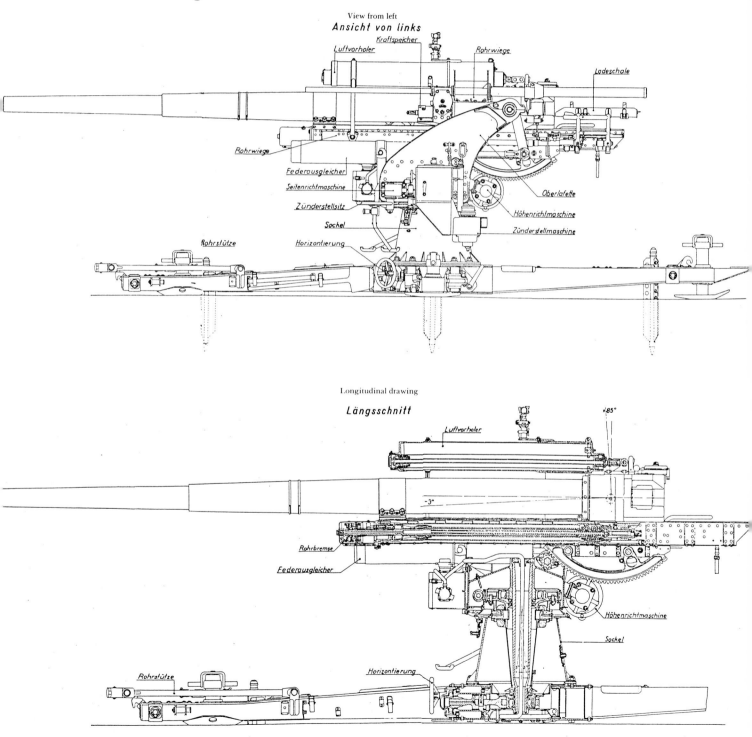

View from left
Ansicht von links

Kraftspeicher
Luftvorholer
Rohrwiege
Ladeschale
Rohrwiege
Federausgleicher
Seitenrichtmaschine
Zünderstellsitz
Sockel
Oberlafette
Höhenrichtmaschine
Zünderstellmaschine
Rohrstütze
Horizontierung

Longitudinal drawing
Längsschnitt

Luftvorholer
+85°
−3°
Rohrbremse
Federausgleicher
Höhenrichtmaschine
Sockel
Rohrstütze
Horizontierung

The picture at left shows the 88mm Flak 18 from the rear. The octagonal bed plate is screwed to the cross mount with the upper mount. On the left side of the upper mount is the fuze setter with two fuze setting mouths.

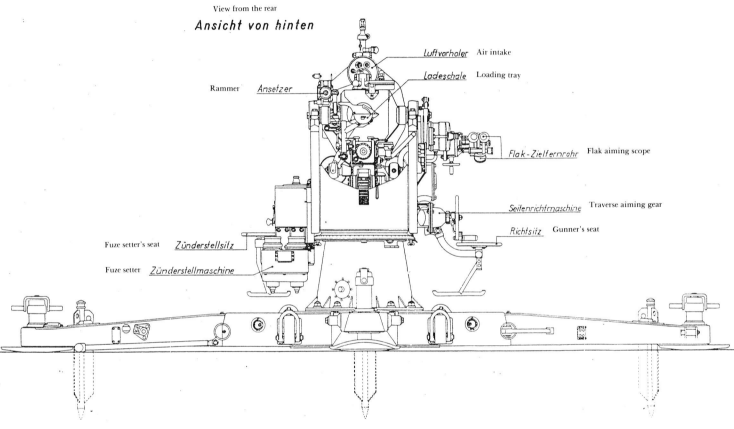

View from the rear

Ansicht von hinten

Luftvorholer — Air intake
Ladeschale — Loading tray
Rammer — *Ansetzer*
Flak-Zielfernrohr — Flak aiming scope
Seitenrichtmaschine — Traverse aiming gear
Richtsitz — Gunner's seat
Fuze setter's seat — *Zünderstellsitz*
Fuze setter — *Zünderstellmaschine*

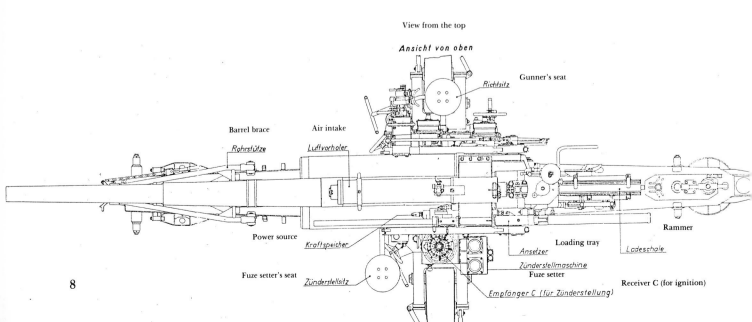

View from the top

Ansicht von oben

Richtsitz — Gunner's seat
Barrel brace — Air intake
Rohrstütze — *Luftvorholer*
Power source — *Kraftspeicher*
Fuze setter's seat — *Zünderstellsitz*
Empfänger C (für Zünderstellung)
Anselzer — Loading tray — Rammer
Zünderstellmaschine — *Ladeschale*
Fuze setter
Receiver C (for ignition)

The crew of an 88mm Flak 18 in a Henschel truck with their gun, going cross-country. Note the brake line with which a crewman had to brake the gun trailer from the truck.

Getting out during a pause in the march was not easy, on account of the truck's high sides.(BA)

This 88mm Flak 18 is towed by an 8-ton tractor, Kfz. 7 Zgkw. 8t Type "HLm 10." The crew jumps from the truck to put the gun into position.(BA)

In 1936 the first improved 88mm Flak 36 guns reached the troops. The mounting platform was now rectangular. The gun itself consisted of a three-section inner barrel (loading chamber, rear and forward sections), held together by a double jacket and a tension ring. The front section was a separate piece so that only the worn part could be changed, as the parts of the barrel did not all wear at the same rate. No particularly long lathes or borers were needed during manufacturing either. After steel covers were introduced in order to avoid expensive brass ones, and with the use of steel rotating bands on the shells, single-piece barrels were made again as of 1943. Ballistically as well as dimensionally, the barrel of the 88mm Flak 36 was identical to that of the 88mm Flak 18, so that they were interchangeable.

The front and rear spars of the Type 36 were now identical. The two limbers of the 202 trailer both carried dual wheels now, so that they too were interchangeable, so the gun could be towed in two positions. The fuze setter, now firmly mounted to the gun, had two openings. The elevation and traverse aiming gear now had a faster speed. Since the springs in the equalizer were stronger now, the elevation aiming gear could be used more easily.

Just a year later, in 1937, an improved gun with the designation 88mm FLAK 37 reached the troops. Its improved direction indicator system required the gunners to keep the pointers, which were controlled by the fire control device via the 108-strand remote-control cable, in line. (Further technical data of the 88mm 18, 36 and 37 can be found in the tables in the appendix.)

Barrel of the 88mm Flak 18.

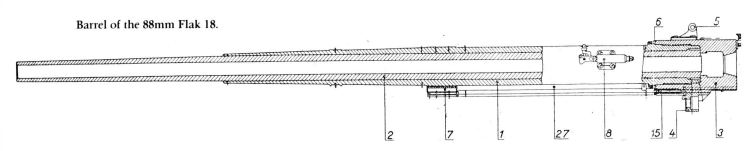

The barrel consists of the jacket (1), the inner barrel (2) installed to allow play but with support, and replaceable as well, the easily detachable breech (3) with barrel brace (4) and pillow block (5), the pressure ring (6) and the removable barrel guide (7). The catch (8) for the rammer was attached to the jacket, the counterweight (9) for the closing spring of the breechblock to the base. (From L.Dv. 436)

Below: The barrel of the 88mm Flak 36 consisted of the inner barrel (1) with rear removable liner (8), forward removable liner (9), forward inner barrel (10), rear inner barrel (11), the spring nut (12) and pressure ring (13). The bayonet ring (5) united the gun jacket (2) with the inner barrel (1). The base (3) was firmly attached to the jacket (2) by the tension screw (4). The barrel guides (6) served to position the barrel in the cradle. The protective sheath (7) was there to protect the course of the cradle. (From the supplement to L.Dv. 436)

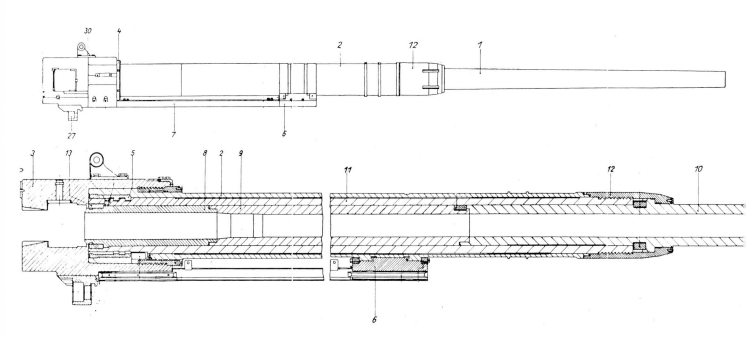

The barrel construction was not restrictive in terms of the gun's designation, for the 88mm Flak 18 barrel was ballistically and dimensionally identical to that of the 88mm Flak 36 and the two were interchangeable. The features that made the two gun types different were found in the outriggers.

Outriggers of the 88mm Flak 18. The front spar (1), rear spar (2), one side spar (4), octagonal sheet-metal plate (5), trailbox (10), two bearings (11) for the barrel braces.

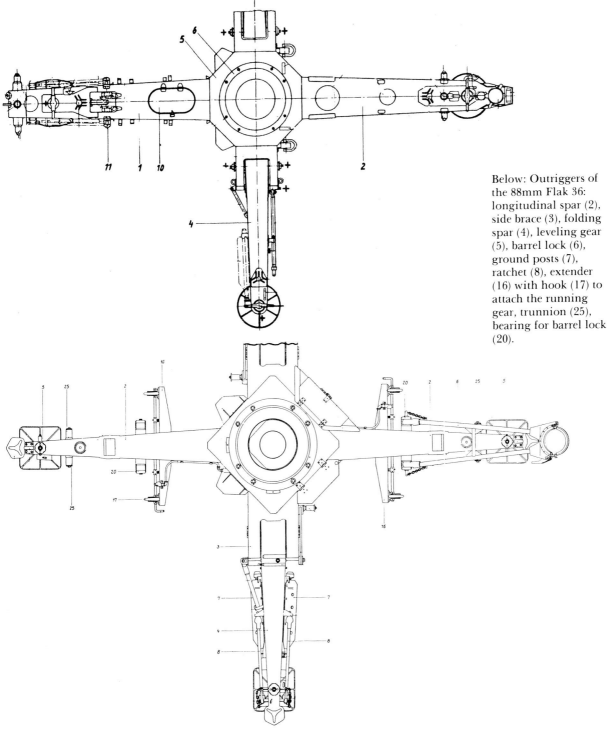

Below: Outriggers of the 88mm Flak 36: longitudinal spar (2), side brace (3), folding spar (4), leveling gear (5), barrel lock (6), ground posts (7), ratchet (8), extender (16) with hook (17) to attach the running gear, trunnion (25), bearing for barrel lock (20).

Final assembly of the 88mm Flak 36.(BA) Note particularly the elevation and traverse aiming machinery. With the help of the left handwheel, the gun was adjusted vertically; with the right handwheel, horizontally. When aiming via the fire control device, the shot data were transmitted through a 108-strand remote-control cable to the lamp indicators over the handwheels. The lamps, lighting in three concentric circles, had to be brought into line by the gunner with the help of the hand wheels. For elevation adjustment, one lamp position in the outer circle was equal to 1/10 degree, in the central circle 1 degree and in the inner circle 10 degrees. For traverse adjustment, the full circle was divided into 6400 segments. One lamp position in the outer circle equaled 6.4 sections, in the central 64 and in the inner circle 640 sections.

Below: This 88mm gun is equipped with the Transmission Device 36. Instead of lamps, pointers had to be lined up.

The fuze setting data were also transmitted by the lamps or, as above, the pointers, with the fuze setting given in degrees from a cross. We see the fuze setter's seat in the picture above, and behind it the box with the command drive. The indicator was controlled by the handwheel and the data transmitted to the mouth of the fuze setter. There the shells to be fired are fitted with fuzes. The mechanical power for this came from the fuze setting drive which was operated by a crank.

Above: Ammunition movers pass shells on and have already placed two in the fuze setting mouths. The circumference of the fuze cap was marked off in 360 degrees. Formerly a cross was placed at zero degrees, hence the slang expression "straight from the cross." Turning the fuze cap 90 degrees, for example, gave a fuze position of 90 degrees from the cross.

In the picture at left, the fuze is set at sixty degrees by hand through the use of a wrench. That was done for safety reasons and also shortened the time needed for fuze setting by stopping the setter's operating time.

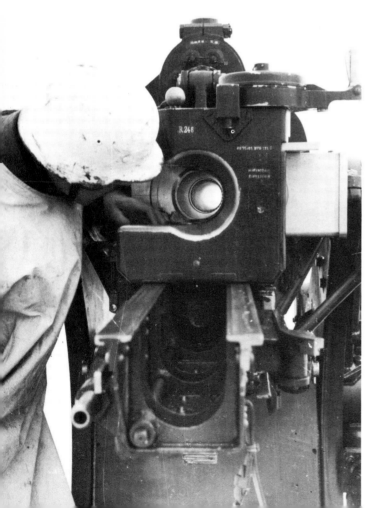

Upper left: The loader, the K3, has taken a shell out of the fuze setter and is loading it into the 88mm gun.(BA)

Upper right: To avoid injuries, every loader had to wear a leather loading glove on his right hand. In the picture, the K3 is loading the gun, and the sliding rails on which the barrel recoiled when firing can be seen. On the base are the spring housings with a handle for opening the latch manually, the handle of the ejector, and the securing handle.(BA)

At left we look through the opened breech into the inner barrel with its 32 rifling grooves (1.05 mm deep and 5.04 mm wide). The spring housing of the shearing handle, with its handle and latch, is on the base. The locking wedge sticks out of the base at the right.(BA)

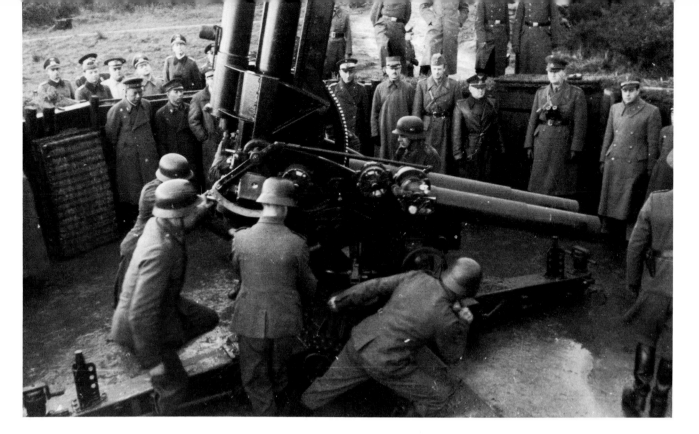

Spring equalizer

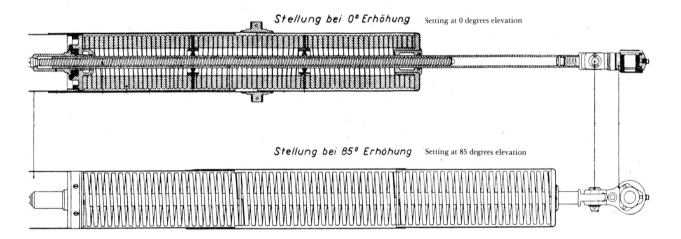

Stellung bei 0° Erhöhung Setting at 0 degrees elevation

Stellung bei 85° Erhöhung Setting at 85 degrees elevation

Above: View of an 88mm Flak 18 demonstration for foreign officers. The gun crew is just changing aim; the traverse gunner, the K2, is swinging the gun around while the aiming gear is disengaged. Note the great elevation of the barrel, with the spring equalizer sticking far out.(BA) (See also the middle drawing.)

Right: The outer spring tube of the left spring equalizer can be seen in the front, with its rod engaging the toothed arc.(BA)

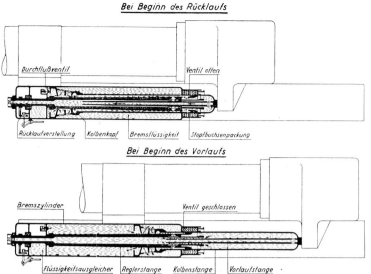

At beginning of recoil

Wirkungsweise der Rohrbremse
Bei Beginn des Rücklaufs

Durchflußventil   Ventil offen

Rücklaufverstellung   Kolbenkopf   Bremsflüssigkeit   Stopfbuchsenpackung

Bei Beginn des Vorlaufs

Bremszylinder   Ventil geschlossen

Flüssigkeitsausgleicher   Reglerstange   Kolbenstange   Vorlaufstange

At left, the inner spring tube of the spring equalizer can again be seen sticking out. Note particularly the recoil brake under the barrel. It had the job of limiting the recoil of the barrel and regulating the counter-recoil.(BA)

Through the cooperation of 10.7 liters of brake fluid (including 1 liter in the fluid equalizer), jets, pistons and valves, the recoil energy of the barrel was gradually absorbed. (See also drawing above.)

The advance of the barrel took place under the influence of the recuperator on the barrel. At the lower left one sees the piston rod over the recoiled barrel after firing; it linked the barrel to the recuperator piston.(BA)

The contents of the recuperator included 39 liters of brake fluid and 19.8 liters of air, compressed to between 38 and 39 atmospheres.

Wirkungsweise des Luftvorholers.

Operation of the Pneumatic Recuperator

Resting position
Ruhelage   Kolben
Druckluft
Bremsflüssigkeit

Recoil
Rücklauf   Bremsflüssigkeit steigt im Luftbehälter
Druckluft wird zusammengedrückt.
Vorholerkolben vom zurücklaufenden Rohr zurückgezogen. Bremsflüssigkeit durch Durchflußkanal in Luftbehälter gedrückt.

Counter-recoil
Vorlauf   Zusammengedrückte Druckluft entspannt sich und drückt auf die Bremsflüssigkeit.
Bremsflüssigkeit wird durch Durchflußkanal in den Vorholzylinder gepreßt, drückt gegen den Vorholerkolben und schiebt ihn vor, wodurch Rohr mit vorgezogen wird.

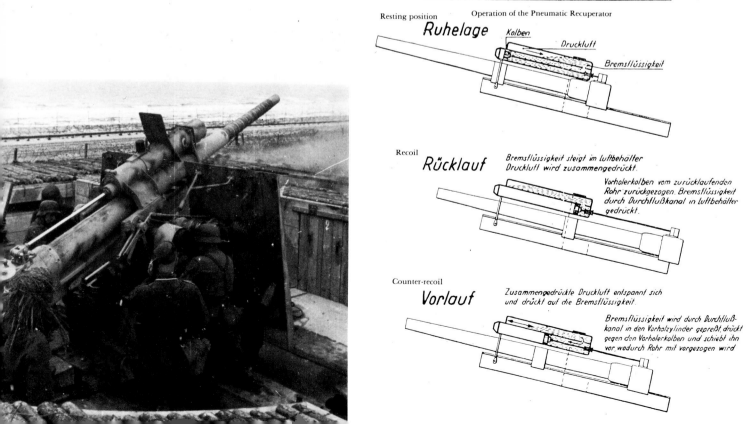

Loading and aiming drill with an 88mm Flak.

Below: This 88mm Flak 18 is set up on a firing range to test its so-called relative power. The shell passes through frames set up at specific intervals, with wires stretched across them. When the shell touched the wire, it was registered automatically. From the time that the shell required to go from one frame to the next, the muzzle velocity, the Vo, of the gun was calculated.

This is how they paraded before Hitler's reviewing stand . . .

. . . and this is how they rode through Russia at his command. (2 x BA)

The 88mm Flak was originally intended for defense against air targets. But even in the Spanish Civil War, the anti-aircraft guns taken along by the volunteer *Legion Condor* were used with success in place of lacking artillery against tanks, bunkers and pinpoint targets. In World War II as well, the 88mm Flak was put to a great variety of uses on account of its great mobility, plus its rapid rate of fire (15 to 25 rounds per minute, depending on the crew's level of training).

In Germany the population and the military supply centers were to be protected against air attack. Naturally, only the most vital points could be covered, and many cities and objectives remained of necessity unprotected. At the front, the Flak guns were used for many purposes besides protecting the troops from air attack, such as action against tanks and bunkers and support of ground troops. For front use, the 88mm Flak was equipped with a shield in 1940. On the coasts, air and sea targets had to be engaged and attempts to land had to be repulsed.

Action in France in the summer of 1940.(BA)

In this 88mm Flak battery on the Atlantic, the guns are set up so that at least three of them have a free field of fire at the sea, in order to be able to engage ground and sea targets during attempts to land.

Above: Action on the
Channel coast in the
summer of 1940.(BA)

Right: The "Caesar" gun of
an 88mm Flak 36 battery is
moved by manpower.(BA)

An 88mm battery secures an
airfield from which Stuka
planes are just taking off on
a combat mission.(BA)

Above: The gun crew of an 88mm Flak 18 being drilled. To the right of the gun barrel is the lamp indicator for the fuze range.(BA)

Below: The "Bertha" gun of a very successful 88 battery. Each ring on the gun barrel indicates participation in shooting down an enemy plane.(BA)

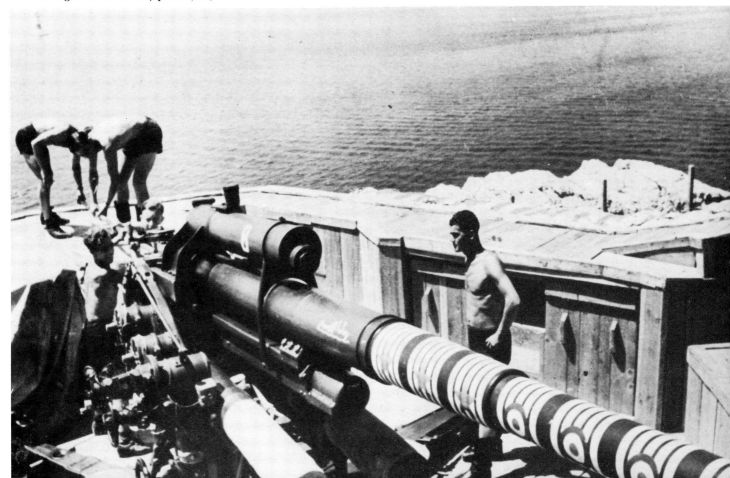

Since the 88mm Flak was also used to support the troops in land action, the guns were very soon supplied with a shield. The picture above shows an 88mm Flak 18 with such a shield. This gun, though, is in a position made exclusively for anti-aircraft action.(BA)

Left: On the right side of the shield was a flap that could be opened by the K2 to use the aiming telescope during land action. The white winter camouflage paint with which the guns and vehicles were covered, especially in Russia, is **flaking off the shield here.(BA)**

Above: This 88mm Flak is being towed to the front by an 8-ton towing tractor (Sd.Kfz. 7).(BA)

Left: An 88mm Flak 18 with a shield in action.(BA)

Right: A pause in the battle is used for personal hygiene.(BA)

Upper left: An 88mm Flak on the Atlantic coast. The gun bears light camouflage paint to blend with the color of the sand.

Upper right: The "Anton" gun of an 88mm Flak battery has been put in position on the shore of a bay to protect seaplanes landing there. The wicker ammunition baskets each held three 88 shells.(BA)

Left: An 88mm Flak 18 gun is pulled across a pontoon bridge in France in 1940 by manpower.

On the western front in 1940 the 88mm Flak was used in large numbers to attack tank turrets and concrete bunkers on the Maginot Line. The 88mm antitank shells penetrated many tank turrets and concrete walls, detonated inside them and put their crews out of action. It was most effective to fire on bunkers from a range of 600 to 2000 meters. In the upper picture we see an 88mm gun firing on a bunker at the upper Rhine crossing near Breisach.(BA) The picture below shows the citadel at Boulogne after being fired on by antitank shells from 88mm Flak guns.(BA)

These varied combat tasks required types of ammunition of which only the most commonly used can be mentioned here. The antipersonnel shell, with a fragmenting load of 870 grams of TNT or Amatol and a time fuze, had the effect of shrapnel. But when they were used against air targets, they proved to be less effective than had been assumed before the war. At a range of 30 meters the target could not be destroyed; only at a range of about 9 meters could it be shot down. All attempts to improve the fragmentation effect by the war's end had little success.

Explosive shells with impact fuzes were used against hidden targets. Shells with delayed fuzes, with an angle of descent below 20 degrees, detonated only 50 meters beyond the point of impact.

Antitank shells of hardened steel and hollow-charge shells were very effective against all tanks then in service as well as many concrete structures. More and more often, the 88mm Flak troops were therefore used in assault points. Such "fire service" action often resulted in heavy losses of well-trained soldiers, who had to be replaced from home guard units. The result was that the vacancies there often could be filled only by students and Hitler Youth members as Luftwaffe helpers (Flak helpers), or by Reichsarbeitsdienst (RAD) workers, or even young people from the occupied territories. Even female Flak helpers were used, especially with floodlights, fire-control devices and for aircraft spotting.

The most frequently used types of ammunition were, in brief:

88mm anti-personnel shell: L4.5 cartridge with nose fuze AZ 23/28 or time fuse Zt/Z S/3 for Flak 18, 36, 37.

88mm antitank shell: Cartridge 40 (AP) for 88mm Flak 36, 41.

88mm antitank shell: Cartridge 39 with base fuze for 88mm Flak 18, 36, 37.

88mm candle shell: Cartridge L 4.4 (a tracer shell with a burning duration of about 23 seconds).

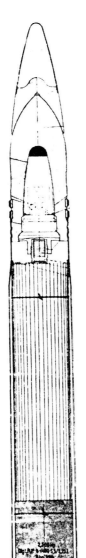

Explosive charge of the 88mm antitank shell (explosive weight 148 grams) Fp u. Mitropenta

Explosive capsule housing Sprengkapsel Pl

Bd Z f 88mm Antitank shell

Nose cone

Cap

Shell body

Closing screw

Liners

Rotating bands (K P S)

Lighting material

Firing material

Tracer trajectory cover

Celluloid disc

38 grams of lead line

Cord binding

Cartridge case (6347) of 88mm Flak 18

Primary charge ca. 2500 kg

Digl RP-B-(460-5.5/2.75)

Cord binding

Cartridge bag

Secondary charge

20 grams Mz. Man. N P (1.5 1.5)

Fuze C/12 n A 4

Cutaway drawing of an antitank cartridge 39.

Below: This is how a French tank turret looked after being fired on by an 88mm Flak with antitank shells.(BA)

Right: Ammunition for
an 88 Flak is supplied
by a ferry.(BA)

An 88mm Flak gun is
prepared for action on a
ferry. Instead of being
secured by shafts in the
ground, the gun is lashed
to the ferry by lines.

Below: This ferry with an
88mm gun is anchored in
the middle of a harbor.

Above: Four guns of an 88 battery are being towed by 8-ton tractors onto a field to take up positions in a square. Each gun is located 70 to 80 paces from the middle of the battery, where the distributor box is located. A 108-strand remote-control cable leads from the distributor to each gun. In the picture at left, they are still rolled on cable drums on the rear fenders.(BA)

These cables transmitted the firing data from the firing director, which is likewise connected to the distributor.

The lower picture shows a built-up position. In order to increase firepower, the number of guns in a battery was increased by war's end to six, twelve and even 24 guns.

The battery leader has chosen the locations of the guns. They are lowered from the limbers and at first camouflaged lightly. The barrel is pointed in the direction of the main combat area. A crewman drives a spike in with a sledge hammer.

After the guns have been placed in position, they have to be adjusted, which means the guns must have the same zero setting as the firing director. Adjusting is done by the section leader with the help of a panoramic telescope on the pneumatic recuperator. He is telephonically in touch with the command post via his headset.

A headlight direction indicator is provided to help the aircraft spotter better identify airplane types.(BA)

The English air attacks took place almost exclusively at night. Then the scene over German cities often looked like this one over Hamburg.(BA)

Below: By changing the propellant charge, attempts were made during the war to decrease this great muzzle fire, by which the Flak batteries were easily recognized.

As soon as all section leaders have reported their guns "ready to fire" to the "technical shooter" by telephone and a target has been chosen within range of the battery, the command "Group fire—Group!" is given. The fire bell rings for three seconds and all the guns fire at the same time.

Right: During a long firing duration, the used, empty cartridge cases are, at first, simply thrown out of the gun position to keep them out of the way.

With the rapid rate of fire of 15 to 20 shots per minute, the wear on the gun barrels was very great. In the first two years of the war, barrage fire was often shot at night, as the number of floodlights was not sufficient and the use of radar was just beginning. Thus an ammunition consumption of 200 rounds per gun in one night was not unusual. Many a barrel could not withstand this use, and barrel explosions occurred, not without danger for their crews.

Above: Unloading the 88mm Flak in Tripoli harbor after good crossing turned out to difficult.(BA)

Left: On the way to the fron the 8-ton towing tractor is n bothered by the rocky terrair Only the barker on the rear limber had it hard on dusty roads.(BA)

Below: Not yet ready for action. But the men have already dug themselves in, remembering the saying, "Sweat saves blood."(BA)

Above: The command to take up positions has been received. The gun is lowered. The barrel is still attached to the barrel brace.(BA)

The barrel brace is detached and is folded forward onto the spar. While the gun is made ready to fire, the section leader observes the area.(BA)

Despite the stony ground, this gun has been somewhat dug in. The battery officer observes the shot accuracy through a periscope.(BA)

An 88mm Flak 18 in action in North Africa. To maintain a rapid rate of fire, the ammunition movers and loaders have to work particularly hard in the heat. Instead of hot, heavy steel helmets, the crew wears sun helmets.(BA)

A barrel is worn out and must be changed. A maintenance troop has come with a replacement barrel and a crane truck to change barrels on the spot.(BA)

The new barrel is placed on the rails of the gun cradle with slides and pushed forward. The inner barrel, jacket and breech are replaced in one operation.(BA)

Above: After the fighting in Africa ended in 1943, many damaged 88mm guns were left behind on the battlefields.(BA)

The war went on in Italy. The 88mm Flak is seen here in a fortified position near Nettuno.(BA)

The 88mm Flak was also used in street fighting. The picture below was taken during the summer of 1944 in an Italian city (probably Florence).(BA)

On no other front sector were the 8-ton towing tractors as necessary for the 88mm guns as in the east. Whether through mud, ice or snow, the 88 guns got where they were going. (2 x BA)

Above: Two 8-ton tractors are towing a broken-down 88mm Flak over a makeshift bridge.(BA)

Left: Here a towing tractor labors to tow an 88mm Flak 36 through the mud—(Autumn 1943, in the central sector in Russia).

An 88mm army anti-aircraft gun crosses a ford, watched with interest by Russian boys. For identification from the air, the swastika flag is painted on the engine hood of the tractor.(BA)

This 88mm gun has been lowered from its limbers and held in position by stakes in the ground. The limbers were not abandoned, though, as positions might have to be changed quickly. The limbers were not abandoned, though, as positions might have to be changed quickly. At right next to the 88 gun, a 20mm gun is in position.(BA)

A so-called "working gun" has been towed onto the airstrip to put down a nest of resistance. Only the side spars have been folded out and braced with boards. As a rule, shooting from a trailer was not allowed because of the strong recoil. But in combat, the situation often required it when there was no time to lower the gun. At such times the gun was fired directly from the limber in the direction it was pointing.(BA)

On this flat terrain the gun has to be well camouflaged so as not to stand out clearly. The ammunition is in baskets close at hand. The gun crew tensely watches from cover.

Above: Scarcely has the battery taken position on a field when the air raid signal is heard.

Here the 88 gun has been brought up to cover a tank attack on a Russian village near Woronesh. The gun crew observes the action of the tanks on the horizon. The K2 sits in his aiming seat on the gun, which has not yet been lowered.

This gun crew, ready to fire, also watches the burning village tensely from their position.(BA)

Because of frequent position changes, there was often no time to dig the guns in. Thus it was necessary to camouflage them well. Here we see two examples of well-camouflaged 88mm Flak. The guns in the upper picture are in a wheatfield, ready for use against land targets.(BA)

The gun at left was camouflaged to avoid being seen from the air. It is raised to its highest position so as to make at little shadow as possible when the sun is high.(BA)

The "Bertha" (above) and "Caesar" (center, BA) guns, recognizable by the letters on their recuperators, are in action against land targets. "Bertha" has already fired and the barrel has recoiled. The barrel of "Caesar" has already advanced after firing. The shell casing has just been ejected. In the picture below, the gun's scores are shown on the barrel.

The situation on the eastern front has grown much worse. Supplies were needed quickly. Here the difficult loading of an 88mm Flak 18 into a four-engine Me 323 "Gigant" is shown. Men of the ground crew stand by with blocks and one man operates the gun's hand brake to prevent it from rolling back.(BA)

Left: This 88mm Flak 36 stands ready to fire on a straight road leading out of Budapest. The barrel brace is disengaged, the side spars are folded down, but the gun is still on the limbers.

An 88mm Flak 36 gun passes a column of Russian war prisoners on an airstrip.(BA)

The Flak Combat Badge, at left, was introduced on January 10, 1941 for members of the anti-aircraft artillery, and awarded for achievement in firing on air or ground targets. After the introduction of the Luftwaffe's Ground Combat Badge on March 31, 1942 the Flak badge was awarded only for success against air targets. Awards were made according to a point system, with 16 points required for the award. On July 18, 1941 the Army Flak Emblem (right) was introduced. It resembled the Flak Combat Badge except for the different national eagle emblem. Both medals were made of silver alloy.

An 88mm Flak in a wooded position for defensive action on the Düna in 1944.(BA)

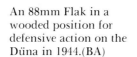

The "Bertha" gun provides protection for an airstrip. The camouflage has been removed to let the gun traverse without hindrance.(BA)

Behind a low snow wall, which was more camouflage than protection, stands an 88 gun crew in white snow suits, fighting against Russian tanks.(BA)

A Russian tank hit by 88mm Flak fire.(BA)

Because of their superb ability to penetrate armor plate, the 88mm guns were used against ground targets more than against aircraft on the Russian front. They were used by Flak combat troops or as so-called "working guns" by the troops on the most advanced fronts. The picture above shows a Flak combat troop's gun in the Nevel area in 1944.(BA)

Left: An 88mm Flak 36 rolls past an airfield and into action.(BA)

Below: The fuze range is set with a fuze wrench before use against ground targets.(BA) The gun crew wears their winter clothing with the white side out. After the snow melted, they could be reversed.

Here are four pictures of 88mm Flak guns in winter action in Russia. Note the unusual winter camouflage in the picture at the right center. (1 x BA)

Above: An 88mm Flak battery stops during an advance near Gauri, Estonia in 1941.

Left: A Flak combat troop in action against Russian tanks on the "Burma Road" near Volkov in January of 1943.

Below: Ammunition boxes are stacked beside this 88mm gun at an airfield in Russia. Each basket holds three 88mm shells ready for use in the gun.

Anti-aircraft guns were also mounted on railroad cars. They could thus be moved quickly from one place of action to another, though they were limited to railroad tracks. An 88mm gun is seen above mounted on a special railroad car. The car is supported underneath by hydraulic jacks and iron braces to stabilize the gun better when firing.(BA)

In the two lower pictures one sees that the side spars of the guns have been removed. The sides of the railroad car could be folded out to extend the platform. (1 x BA)

Here we see several interesting types of railroad gun cars. In the two upper pictures, the 88mm Flak is on an armored train made up of six armored infantry and gun cars. The crew of about 120 men was also equipped with machine guns and grenade launchers. On the four-axle flatcar at right, an improvised wooden gun platform has been built. On the stake flatcar at the lower left is an 88mm Flak 37 loaded for transportation. Note that the folded side spars, which were removed from railroad-car guns, are present. An unusual type of gun car is seen in the lower right picture. On this car are two 88mm guns, with the ammunition in the center of the car.

The 88mm Flak was often used as an armor-piercing weapon at assault points, which led to considerable losses. Better armoring of the vehicle and gun, as well as increased mobility, was to decrease losses. At right, an 88mm army Flak 18 is towed by an armored 8.ton tractor.(BA)

To avoid the loss of time in detaching and attaching limbers, the guns were mounted on self-propelled 8- or 12-ton tractors. The picture below shows an 88mm Flak on Sd. Kfz. 9, used along with tanks in an attack.(BA)

The 88mm Flak on an 8-ton tractor.

A
88
b:

T
tl
K
w

Sc
oi
la
P
Sc
o"
"Sc
oi
pa
Tl
th
p
b:

Above: An 88mm Flak gun
on a 12-ton tractor (Sd.Kfz. 8)
with an armored cab.

Left: This 12-ton army
tractor with an 88mm Flak
gun takes a risky trip over a
makeshift bridge. "Will it
hold or not?" The men watch
nervously from the shore. A
tank tried to go around the
bridge but, it seems, got stuck
in a bad spot.

Below: The 88mm Flak 18
mounted on a 12-ton tractor
(Sd.Kfz. 8) as a Panzerjäger,
with armored grille and cab.

R
tl

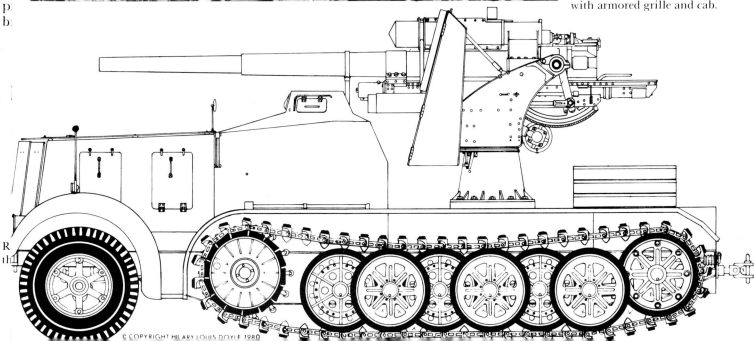

© COPYRIGHT HILARY LOUIS DOYLE 1980

The 88mm Flak 41 on a special vehicle.(BA)

The drawing below shows a test vehicle with the new 88mm Flak 41.

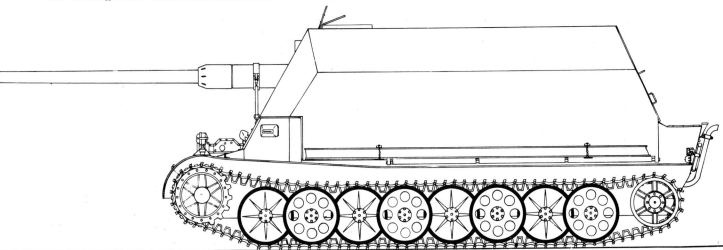

This 88mm Flak 41 with Panther I components was also a test vehicle.

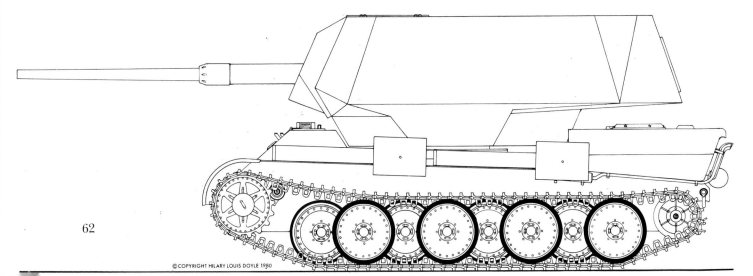

62

© COPYRIGHT HILARY LOUIS DOYLE 1980

Above: An 88mm Flak gun on a 12-ton tractor (Sd.Kfz. 8) with an armored cab.

Left: This 12-ton army tractor with an 88mm Flak gun takes a risky trip over a makeshift bridge. "Will it hold or not?" The men watch nervously from the shore. A tank tried to go around the bridge but, it seems, got stuck in a bad spot.

Below: The 88mm Flak 18 mounted on a 12-ton tractor (Sd.Kfz. 8) as a Panzerjäger, with armored grille and cab.

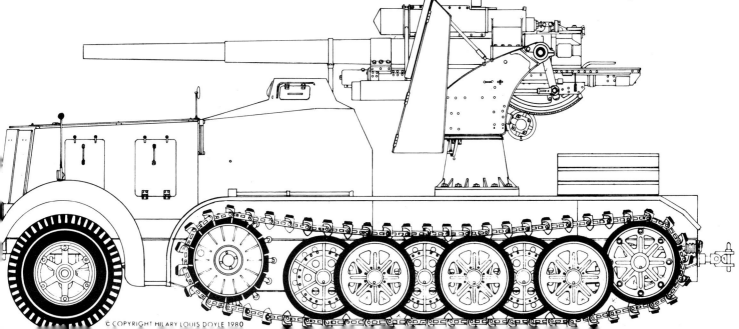

© COPYRIGHT HILARY LOUIS DOYLE 1980

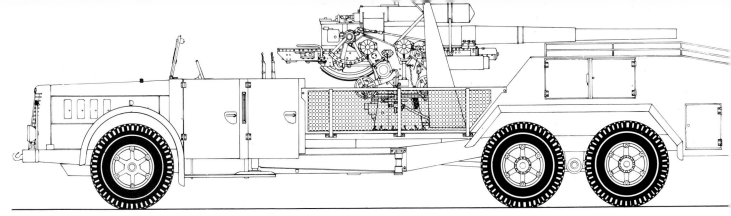

This drawing shows an 88mm Flak 18 on a special Büssing-NAG vehicle. It is probably a prototype that never went into production.

The 88mm gun was such an outstanding armor-piercing weapon that in 1944 thirteen antitank artillery units were equipped with it. In somewhat modified form, the 88 was also mounted on the chassis of the Panzer III and later the IV to make the gun more mobile in rough terrain. These Panzerjäger vehicles were known by the names of Hornisse (Hornet) and Nashorn (Rhinoceros). Later the Battle Tank V (called "Jagd-Panther" (Hunting Panther), and the VI, called "Elefant", were also equipped with the 88mm gun.

Two interesting pictures and a drawing of the 88mm Flak on an 18-ton tractor (Sd.Kfz. 9) with armored grille and cab.

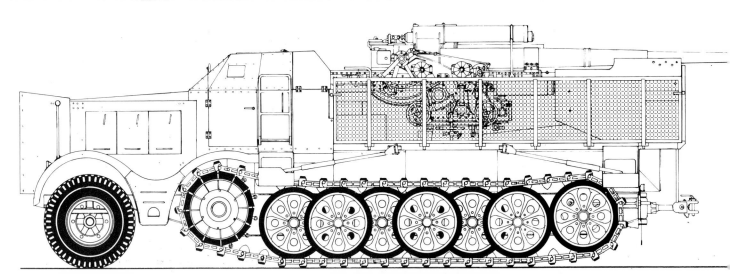

Mounting the 88mm Flak on a towing tractor was not satisfactory, since there was no lateral support when in action. Thus a special vehicle was developed for the 88mm Flak 37. This vehicle, though, did not go into series production as a Flakpanzer.

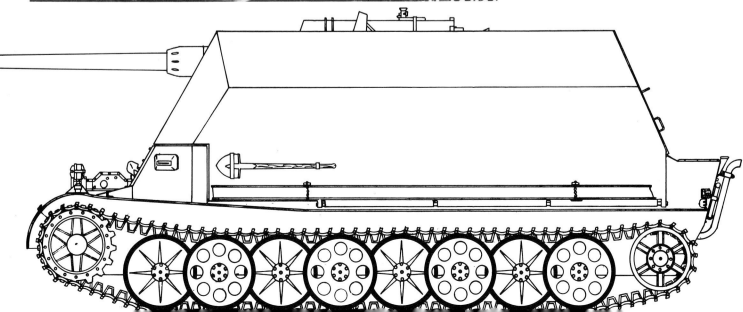

Further interesting models were these prototypes mounted with 88mm Flak 41. They did not go into production either. But the anti-aircraft gun, thanks to its good ballistic qualities, did become the most effective antitank gun. Attempts were made to improve them more and more for use on the battlefield without giving up their original purpose as anti-aircraft guns. This led to a few failures, until the 88 was finally mounted, after a few modifications, on an armored vehicle as a pure Kampfwagenkanone (KwK).

Above and right: A "Nashorn" with an 88mm gun. The muzzle brake and barrel brace are easy to see.

To indicate the course by which the 88mm Flak became the 88 mm KwK, the most important types will be listed here:

Sd.Kfz. 164/1 88mm Pak 43/1 L/71 on Panzerjäger III as "Hornisse", later Sd.Kfz. 161 as "Nashorn" on Panzer IV chassis;
Sd.Kfz. 184 88mm Pak 43/2 L/71 on Panzerjäger Tiger P as "Elefant";
Sd.Kfz. 173 88mm Pak 43/3 L/71 on Panzerjäger Panther (Jagdpanther).
The most important change from the 88mm Flak to the 88mm Pak probably consisted of the muzzle brake and the electric firing.

Right: An "Elefant" (Sd.Kfz. 184) with the reliable 88 KwK.

The Jagdpanther (Sd.Kfz. 173) with the 88mm Pak 43/3, L 71 was one of the best of its kind in World War II.

The Tiger I also used the 88 KwK 36 L/56. This gun, developed from the 88mm Flak, had a flat trajectory thanks to its high muzzle velocity (Vo) of 810 meters per second. The 88 KwK 43 L/71 developed later even had a Vo of 1120 meters per second.

Below: This picture shows a Jagdpanther with the barrel of the 88 KwK with a muzzle brake; despite its length, it needed no barrel brace.

Other than the Luftwaffe, the Army, the Navy and some SS units had their own anti-aircraft units, whose heavy batteries were equipped with the 88mm Flak.

During the course of the war, more and more batteries were set up in permanent positions. This proved to be a big mistake. When enemy troops moved onto German territory, many batteries had to be abandoned and blown up, since it was not possible to move them. In addition, no material was saved by setting up permanent positions, as a permanent position usually used more steel than a cross mount.

In closing, the quantities of ammunition carried with the 88mm Flak 18/36/37 by the anti-aircraft artillery in February 1945 will be listed, the data received from the Quartermaster General of the Luftwaffe General Staff:

88mm Flak 18/36/37 (mobile) 3590 rounds
88mm Flak 36/37 (railway) 41 rounds
88mm Flak 18/36/37 (makeshift-mobile) 1493 rounds
88mm Flak 18/36/37 (permanent) 4178 rounds

The end of many 88mm guns.

88mm FLAK 41

Even before World War II it became clear through the fu··ter development of the German Luftwaffe that the performance of the 88mm Flak 18/36/37 soon would no longer suffice to fight against no less progressive enemy planes successfully. Rheinmetall-Borsig thus began in 1939 to develop a gun that would be far superior to the old 88mm Flak. In 1942 the 44-gun 0 Series of this new 88mm FLAK 41 was sent directly to Rommel's troops in Africa, at Hitler's command. 50% of the guns were lost in the Mediterranean during the crossing. In the rest, which were put into service, the usual "teething troubles" could be overcome only with difficulty. Nevertheless, the 88mm Flak 41 was the best anti-aircraft gun in the world at that time, but its production proceeded very slowly. Thus in 1945 there were only 279 of the guns in service.

The 88mm Flak 41 with its swiveling mount stood on cross spars. The five-section barrel construction with its three-section inner barrel for a sheet metal sheath proved to be impractical. From the 153rd gun on, a four-section barrel for a tempered steel sheath was used, and in the end a one-piece barrel for an untempered steel sheath. The breechblock was a semi-automatic shearing-handle type with a hydraulic-pneumatic roller rammer. For use at the front, this gun was equipped with a shield. Not only because of the meager height of 1250 mm and the high Vo of 1000 meters per second, but also because of the greater ballistic data, it outperformed the other 88mm anti-aircraft weapons by far. (See the table in the Appendix.)

For action against air targets, the elevation range to + 90 degrees was a great advantage. The fire control from Command Post I was done via Direction Indicator System 37. The 88mm Flak was portable on Special Trailer 202. In May of 1943 Krupp produced the first gun of this type on a self-propelled mount.

In order to achieve the ballistic performance of the 88mm Flak 41 with the mount of the 88mm Flak 18, 36 and 37 as well, these 88mm FLAK 37/41 were equipped with a lengthened barrel including a muzzle brake. The fuze setter and the roller rammer were also like those of the Flak 41. Because of the longer barrel, the equalizing springs were strengthened and the air pressure in the recuperator was increased. In transport position on Special Trailer 202, the barrel was drawn back on its cradle. Because of ammunition-supply difficulties for these Flak 37/41 guns, only a few of them saw action.

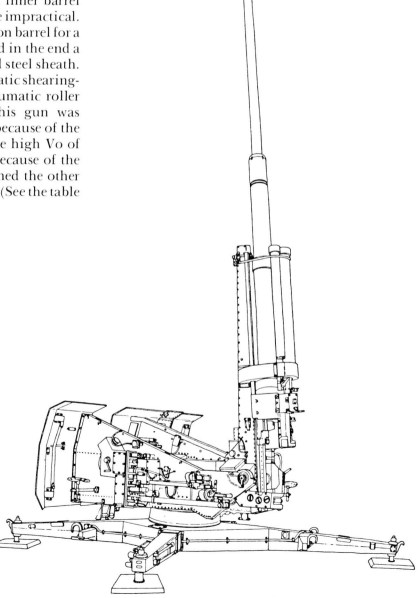

The drawing shows the left side of an 88mm Flak 41 with the fuze setter and a protective shield, as used against ground targets.

An 88mm Flak 41 on the two-axle Special Trailer 202, in transport position. The recoil brake, the recuperator and power source had to be uncoupled and the barrel drawn back. It pointed backward in transit.

Center: The gunners' seats for the elevation and traverse aiming gear were folding against the mount for transit. The barrel of the gun shown here has not yet been drawn back.

Below: The side spars and the shield were folded in when the 88mm Flak 41 was ready for transit.(BA)

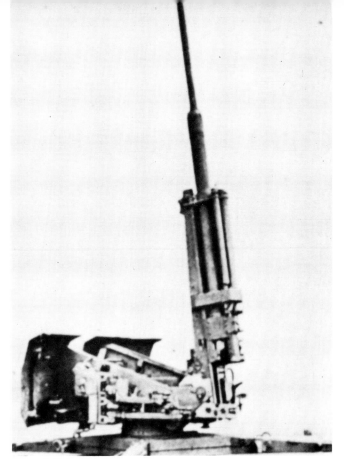

One of the first 88mm Flak 41 with a cross mount. On the left side of the upper mount is the fuze setter with the handwheel for command operation, by which the gun's fuze range settings received from the fire control device were set. With the crank on the side, the mechanical power for setting the fuze was produced.

This 88mm Flak 41 has almost reached its maximum elevation of 90 degrees.

Below: An 88mm Flak 41 in firing position. Directly behind the shield are the two handwheels for the traverse aiming gear and the seat for the K2. Ground targets could be observed through an aiming telescope via the hatch in the shield. Two gunners' seats are attached to the upper mount for the elevation aiming gear. For use against air targets, two elevation gunners were needed, as the 88mm Flak could be aimed only by hand.

The left side of the gun with the fuze setter and the brackets in which the shells were placed by the ammunition movers. The K7 pushed them from there into the fuze setter.

Below: This picture shows an 88mm Flak 41 on a special vehicle with folding side panels.

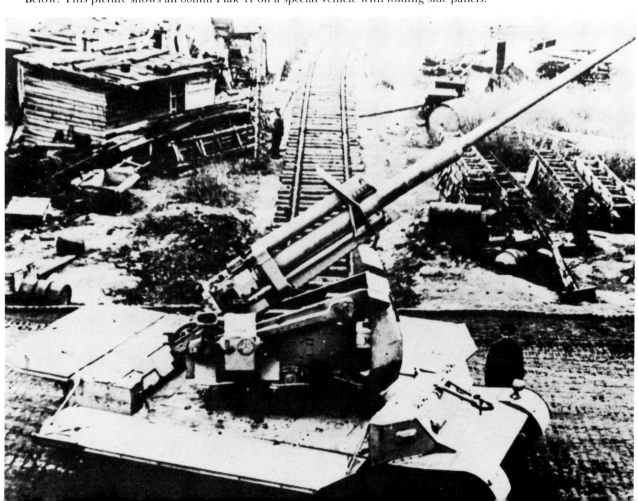

The 88mm Flak 41 on a special vehicle.(BA)

The drawing below shows a test vehicle with the new 88mm Flak 41.

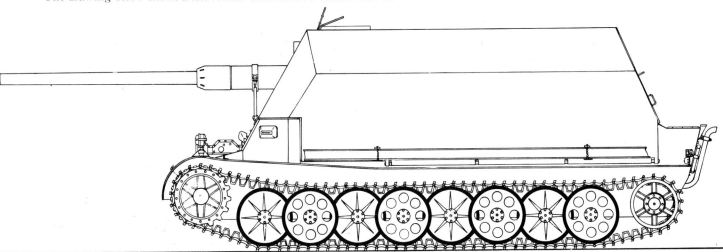

This 88mm Flak 41 with Panther I components was also a test vehicle.

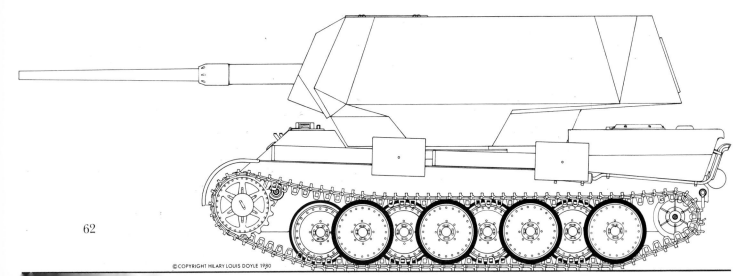

The first 88mm Flak 41 guns were used in Africa. Here, though, the war was over for them in 1943. The two pictures show destroyed guns after the surrender of the *Afrika Korps* in Tunis.

The enemy, of course, had a healthy respect for the penetrating power of the 88mm armor-piercing shells.

The penetrating power of the 88 Antitank Shell 39:

at 100 meters: 194mm
at 500 meters: 177mm
at 1000 meters: 159mm
at 1500 meters: 142mm
at 2000 meters: 127mm

The penetrating power of the 88 Antitank Shell 40:

at 100 meters: 237mm
at 500 meters: 216mm
at 1000 meters: 192mm
at 1500 meters: 171mm
at 2000 meters: 152mm

105mm FLAK 38, 39

The 105mm Flak was originally developed by the Navy and was redesigned by Rheinmetall as the 105mm FLAK 38 for use on land. The gun stood on a pivoting-socket mount. The Pittler-Thoma hydraulic systems of the elevation and traverse aiming gear were driven by direct-current motors, as was the automatic loading system with its two loading mounts. Loading by hand was no longer possible, as the shells weighed 26 kilograms and the shot weighed 15.1 kg. In case of a power failure, the roller rammer could be activated, if necessary, with a cable. The semi-automatic cross-wedge breech with electric firing allowed a rate of fire of 12 to 15 rounds per minute.

The improved 105mm Flak 39 reached the troops in 1940. In this gun, the power supply had been changed from direct to alternating current, in order to take power from the public power lines for permanent batteries. To transmit signals from the fire control system, at first the lamp system was used, and later the Transmission Device 37 with pointers. The original one-piece barrel was changed to a five-piece type, in which the 5.547-meter inner barrel, with 36 riflings, composed of three pieces. As can be seen from the table in the Appendix, the ballistic data were better than those of the 88mm Flak 18/36/37. They were, to be sure, later exceeded by those of the 88mm Flak 41.

The 105mm Flak was used chiefly within Germany, but also in the rear areas of the fronts to protect important places. Some of the batteries were made permanent, others mounted on railroad flatcars. But the majority of them were mobile on the two-axle Special Trailer 203. The gun was unsuitable for front service because of its great weight of 14,600 kilograms in transit form and 10,240 kg in firing position.

According to information from the Quartermaster General of the Luftwaffe General Staff, the shell supply in February 1945 was:

105mm Flak 38/39 (mobile)	898 rounds
105mm Flak 38/39 (railway)	110 rounds
105mm Flak 38/39 (permanent)	876 rounds

The 105mm Flak 38.

4404.38K

The picture above shows the left side of the 105mm Flak 38 gun with the fuze setter, the receiver and motor that went with it, on the upper mount. The fuze range data transmitted via the receiver or otherwise were set by a handwheel on the drive box and transmitted electrically to the positioning gear via the fire control system. If the electric motor failed, the gun could be operated manually by using the crank on the drive box. At left beside the fuze setter in the seat for the K6. The platform for the loader, the K3, could be swung up to 90 degrees for high barrel elevations.

The recoil brake with fluid equalizer, the functioning of which is shown in the drawing at right, was mounted in the barrel cradle between the two spring equalizers. The brake cylinder was filled with 15.5 liters of brake fluid.

Operation of the recoil brake

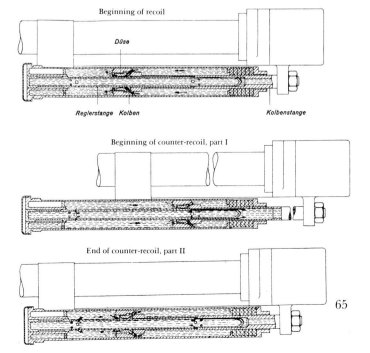

Beginning of recoil

Düse

Reglerstange Kolben   *Kolbenstange*

Beginning of counter-recoil, part I

End of counter-recoil, part II

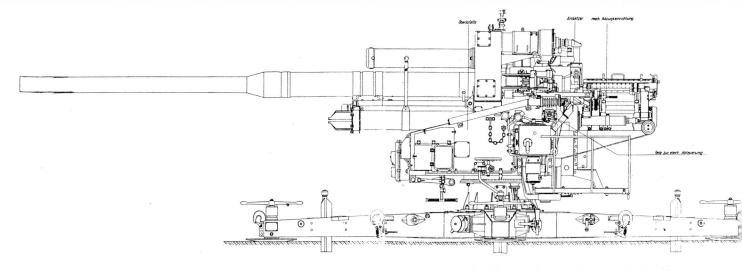

This drawing shows the left side of the 105mm Flak 38 with the already described fuze-setting mechanism, the loading platform and tray, and the electric loading system. The panoramic telescope is mounted in a housing on the recuperator carrying arm, the highest point on the gun.

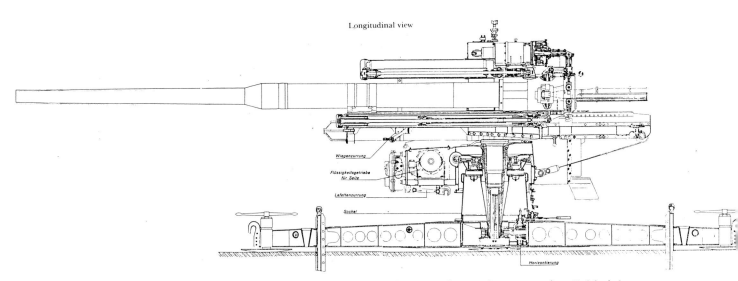

In this longitudinal view of the 105, the recuperator and recoil brake can be seen in cross-section. Behind the recuperator is the rammer motor with its rollers, which drew the cartridge into the barrel.

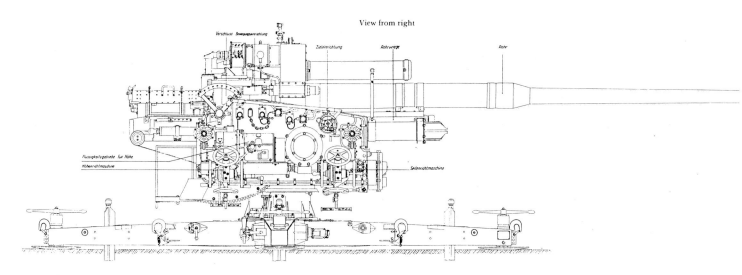

On the right side of the upper mount, the elevation and traverse aiming gear with their handwheels, as well as the gunner's seat, can be seen.

View from above

Seitenrichtsitz   Höhenrichtsitz

Ausgleicher

Zünderstellsitz

Lade-u Stellschale

Ladeschalenantrieb

Plattform

Guns up to number 120 had
a one-piece inner barrel.
1a One-piece inner barrel
2 Barrel jacket
3 Breech
5 Bayonet ring
6 Barrel guide
30 Handle
53 Sliding shoe
118 Opening crank

Gun with divided inner
barrel:
1 Inner barrel
2 Barrel jacket
3 Breech
5 Bayonet ring
6 Barrel guide
7 Protective shield
8 Rear liner
9 Front liner
10 Front inner barrel
11 Rear inner barrel
12 Spring nut
53 Sliding shoe
118 Opening crank with
30 Handle

The two equalizers had the
job of balancing the
forward weight of the
swinging part of the gun at
any elevation. They were
housed in tubes on the
cradle and were elevated
along with the barrel.
275 Screw pressure spring
282 Connecting rod
283 Protective tube
285 Cable to upper mount
bearing

Barrel with one-piece inner barrel

Schnitt A-B

Barrel with divided inner barrel with liners

Schnitt A-B

Equalizer

Ausgleichrohre der Wiege

Stellung bei -3°
wagerecht gezeichnet

Umlenkn
an der W.

Haube an der Wiege

Sicherung

275   282   275   283

Oberlafette

Stellung bei -85°
wagerecht gezeichnet

285

283

283

Installed position

282

Sicherung

The loader stands on the loading platform. The shell lying in the tray before him lies with its point in front of the fuze setting cap. With a short push on the cartridge, the fuze setting cap falls onto the point of the shell and sets the prescribed fuze range. The K3 pulls briefly on the handle of the tray (picture at right), moving the shell out of this tray into the loading tray. This swings in and brings the shell before the barrel. The electrically driven rubber rollers of the rammer move the shell on into the barrel. Then the loading tray automatically swings back out.

A third picture, taken during an inspection of the 1./407 in a position near Düsseldorf in 1941. The two gunners are sitting on their seats, the K2 at the traverse aiming gear in front, the K1 at the elevation aiming gear. With two vertical handwheels each, the gun can be turned horizontally or vertically. With the diagonal handwheels it is motor-driven. The gunner in the foreground stands in front of the switch box by which the electric motors and turned on and off.

A picture for the folks back home. The gunner stands beside the traverse aiming gear. He is still wearing a chest microphone, which the gunners needed for communication with the fire control position. They were soon replaced by headsets.

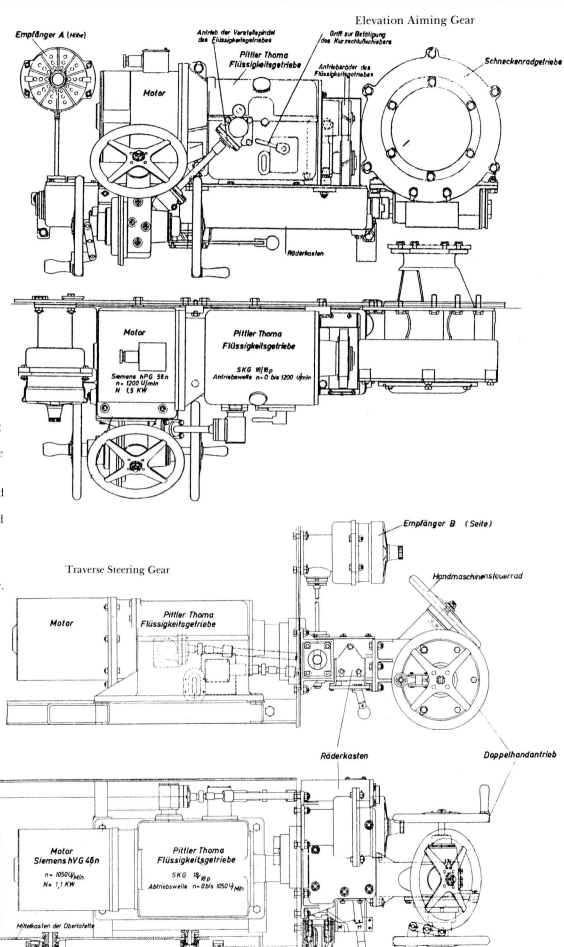

Elevation Aiming Gear

Empfänger A (Höhe)

Motor

Antrieb der Verstellspindel des Flüssigkeitsgetriebes

Pittler Thoma Flüssigkeitsgetriebe

Griff zur Betätigung des Kurzschlußschiebers

Antriebsräder des Flüssigkeitsgetriebes

Schneckenradgetriebe

Räderkasten

Motor

Siemens hPG 56n
n = 1200 U/min
N 1,5 KW

Pittler Thoma
Flüssigkeitsgetriebe

SKG 18/18p
Antriebswelle n-0 bis 1200 u/min

The elevation and traverse steering gear were composed of the motor with Pittler-Thoma fluid drive, the wheel box with dual hand drive and handwheels, the worm drive and the receiver A for elevation and for traverse. The fluid drive for traverse and elevation was switched between the electric motor and the wheel drive. It provided non-stop regulation of the number of turns of the driveshaft at the constant rotating speed of the electric motor.

Empfänger B (Seite)

Handmaschinensteuerrad

Traverse Steering Gear

Motor

Pittler Thoma
Flüssigkeitsgetriebe

Räderkasten

Doppelhandantrieb

Motor
Siemens hVG 46n

n = 1050 U/Min.
N = 1,1 KW

Pittler Thoma
Flüssigkeitsgetriebe

SKG 18/18p
Abtriebswelle n-0 bis 1050 U/Min

Mittelkasten der Oberlafette

Schneckengetriebe

feststehendes Schneckenrad

Kegelräderkasten

69

This 105mm Flak 38 stands in a built-up position in the South. The elevation and traverse gunners receive their data from the fire control device via the lamp receiver. These are shielded from glare effects by leather visors. Later the 105, like the 88mm Flak 38, was fitted with the Transmitter 37 with indicator pointers.

Two impressive pictures of the 105mm Flak in action at night.

The 105mm Flak was seldom used as it is here, in ground combat. Gun "Emil" of a 105 battery on the Oder is firing on a Russian bridgehead.

The picture below shows the crew of a 105mm Flak maintaining their weapon. The ammunition is stacked around the gun, some of it in baskets, the rest in sheet metal sleeves. One shell weighed 26.5 kg, the weight of the shot was 14.8 kg, and the explosive charge consisted of 1.5 kg Amatol and the propellant charge of 5.63 kg of Diglycol gunpowder (Dig.RP).(BA)

Functioning of the Pneumatic Recuperator

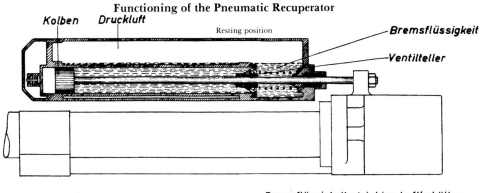

Kolben Druckluft
Resting position
Bremsflüssigkeit
Ventilteller

The contents of the recuperator consisted of 22 liters of brake fluid. The original pressure of the air was 60 atmospheres. At firing, the piston rod was moved by the recoiling barrel. Thus the piston compressed the brake fluid, pushing it through a valve into the air chamber and compressing the air even more. After the recoil ended, the vent plate was pressed against the compression chamber by a spring. The fluid was pushed back into the compression chamber by the compressed air and the barrel was moved forward by the piston rod.

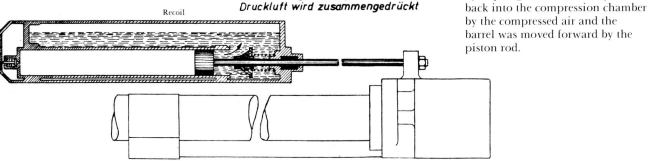

Bremsflüssigkeit steigt im Luftbehälter Druckluft wird zusammengedrückt
Recoil

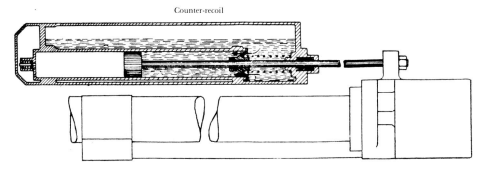

Counter-recoil

A 12-ton towing tractor has pulled a 105mm Flak gun to its position. It is already uncoupled and lowered. One limber of the Special Trailer 203 still stands near the gun.

The B on the recuperator indicates that the barrel is set at 85 degrees.

To be able to move the 105mm Flak faster than on the road, more than 100 of the guns were mounted permanently on special railroad cars. The platform for the gun crew could be widened by folding the sidewalls down. They, like the car itself, were specially braced under the gun. At both ends of the car are containers for ammunition boxes. (2 x BA)

Above are two pictures of a 105mm Flak battery in France late in 1942. The setting of the guns is necessitated by the direction of the railroad tracks. Here a favorable position has been taken on a curve. (2 x BA)

Left: This symbol appeared on the special railroad cars that carried these guns.(BA)

128mm FLAK 40

The first 128mm FLAK 40 guns, like the 105mm Flak, had a cross mount with extenders and two limbers. Because of its great weight, the barrel had to be removed for transport and shipped separately on its own carrier. Since this was very unfavorable for changing an anti-aircraft gun's position, the cross of the pivoting mount was attached to a platform at whose corners four folding horizontal extenders were attached. With a hydraulic lifting jack, the gun could then be lifted onto the two four-wheel limbers of the Special Trailer 220. Even so, the gun remained difficult to move and thus was used almost exclusively in permanent positions or on railroad cars.

The electrically powered aiming gear, the traverse gear having a special high speed, were similar to those of the 105mm Flak. Electric automatic loading was used. The fire control and transmission of the firing data were carried by a remote-control cable from the Direction Indicator System 37.

In 1942 Hanomag delivered the first 128mm twinned Flak 40 anti-aircraft guns. They were permanently positioned in large cities such as Berlin, Hamburg and Vienna, usually on Flak towers. It was possible to move them when they were transferred to the Special Trailer 203 with a transport bridge. The exterior and interior ballistic data were identical to those of the 128mm Flak 40. (See table in Appendix.)

Each of the twin barrels had its own fuze-setting machine, since they could be loaded independently by the automatic electric loaders. A battery of four twinned guns, thus with eight barrels and a rate of fire of 10 to 12 shots per minute, presented a considerable firepower.

The 128mm Flak had a maximum range of 20,900 meters with a maximum altitude of 14,800 meters, a fuze range of 12,800 meters and a shot weight of 26 kilograms, making it the most effective anti-aircraft gun of World War II. Yet as of the end of 1944, its use against enemy battle groups often flying at altitudes of 10,000 meters had only a limited effect.

The development of a 128mm Flak 45 never progressed beyond a prototype by the end of the war. This gun had a lengthened barrel. The shell was fitted with a conical point and weighed 28 kilograms.

The ammunition supply in February 1945, according to information from the Quartermaster General of the Luftwaffe General Staff, amounted to:

128mm Flak twin 40 (permanent)	38 rounds
128mm Flak 40 (mobile)	4 rounds
128mm Flak 40 (permanent)	429 rounds
128mm Flak 40 (railway)	212 rounds

At right, one of the first 128mm Flak 40 guns, still with a cross mount. (BA)

75

The sign on the gun reads:

12.8 cm Flak
Deutsche Wehrmacht
Eingeführt 1942
Gew in Feuerstellung 18000 kg
Schuss pro Minute 10–12
Höchste Schussweite 20900 m
Höchste Schusshöhe 14800 m

This 128mm Flak 40 stands on a platform mount with four folding horizontal extenders.

Here are both versions of the
128mm Flak 40. Above, the first
model standing on a cross mount;
below, the later version with a plate
mount, four folding horizontal
extenders and a loading platform,
which was not available at first.

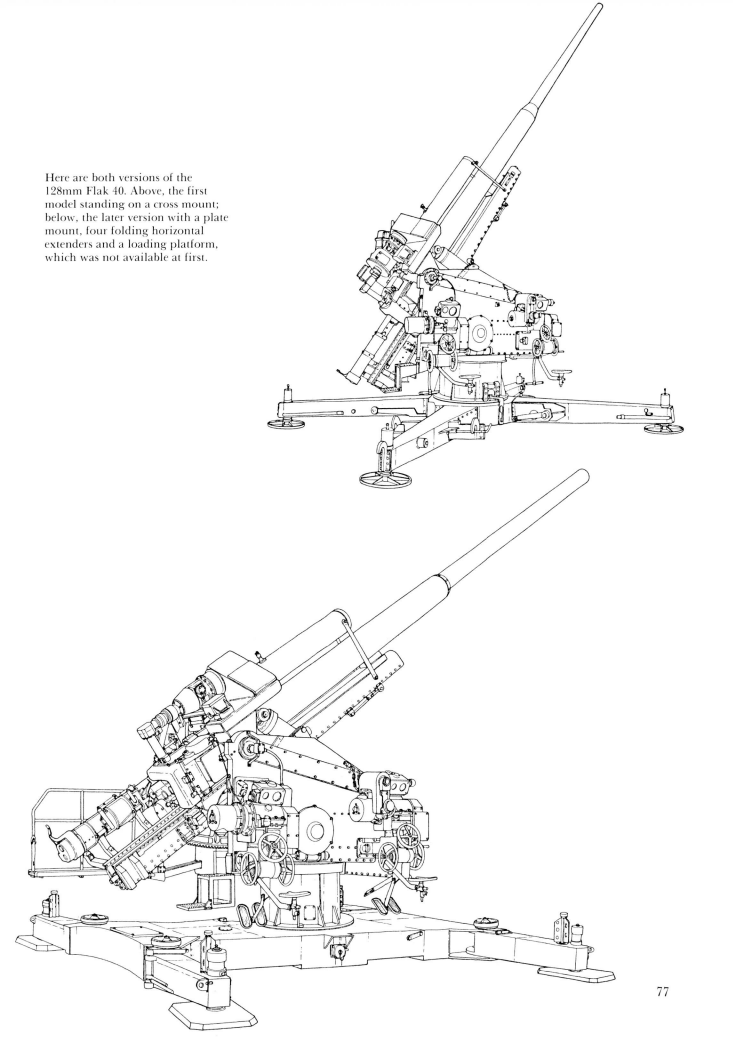

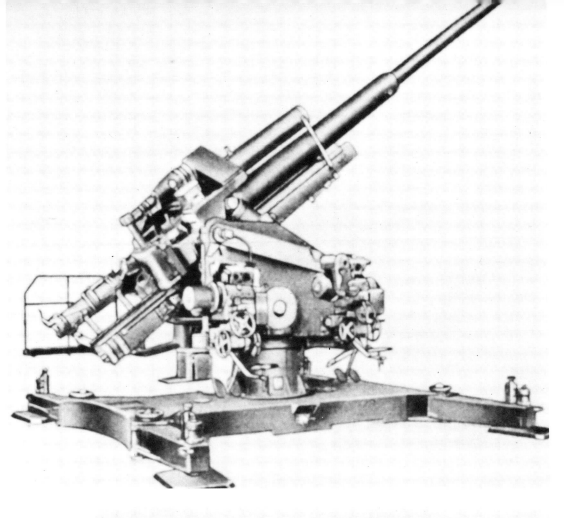

Left: The right side of a 128mm Flak 40 with its elevation and traverse aiming gear, which were similar to those of the 105mm Flak.

The lower pictures show the 128mm Flak 40 on its Special Trailer 220.(BA)

Here the right side of a
128mm Flak 40 can be seen.
In the center of the picture
is the traverse aiming gear
with the traverse gunner's
seat. the K2. Over it is the
direction indicator receiver
for the horizontal firing
data. At left, beside the K2,
sat the K1 elevation gunner,
in front of his elevation
aiming gear and the
direction indicator receiver
for vertical firing data.

This is the front of the 128mm
gun with the switch box for the
electric motors. Two strong
spring equalizers unite the cradle
and upper mount.

On the lower picture we see the
barrel recoil gauge on the gun
cradle near the elevation aiming
gear. This gauge had to be
watched constantly during firing.
The 1300 mark could not be
exceeded; otherwise the firing had
to be aborted.

On the left side of the upper mount is the loading platform for K3 and K4. They were swung out when the angle of elevation was between 60 and 88 degrees. The K3 operated the switches for air and ground firing, the firing button and loading button. In addition, he had to lay the shells, which the K4 gave him, on the placing tray.

On the front of the gun are the main and firing switches as well as the switches for the elevation, traverse, fuze setter, rammer and fuze power receiver, which were operated by the K6. The fuze ranges transmitted from the fire control device were automatically given by the fuze power receiver. Only when these were transmitted by telephone did the K6 have to set the fuze settings by using a handwheel. This can be seen at the left in the upper picture, next to the direction indicator. The K6 stood on a special platform on the fuze setting machine.

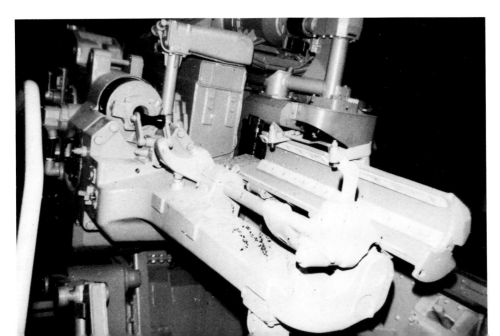

The placing tray is easy to see at the left. When the shell was placed on it, the fuze setting head rose up and placed the fuze. When the loading button was pressed, the shell swung into the loading tray and was pulled into the barrel by the rammer rollers.

The section leader observes the K1 at his work on the
elevation aiming gear. He himself is linked to the
command post by his headset, via a command
line.(BA)

The elevation gunner is seen at his work at the right.
By controlling the aiming speed with the help of a
handwheel, he matches the command indicators of
the receiver, which give the firing data from the
command device, with the direction indicators. A
continuous aiming speed was achieved by a Pittler
fluid drive, as with the 105mm Flak. On the
recuperator above the gun barrel is the letter B,
showing that this is the ''Bertha'' gun, as well as the
number 60, indicating the compressed air in the
recuperator.

Because of the great weight of 18,000 kilograms in firing position and the resulting difficulty of transport, the 128mm Flak was only used in permanent positions. In this picture the 128 is shown on a flak tower. These towers were usually also air raid bunkers for the population.

Below: This picture was taken from the ammunition bunker that extended all the way around the gun position. On the loading platform of the gun stand the K4 and, to the left of him, the K3. The K6 is turning the fuze setter on the direction indicator with a handwheel, which was not necessary under automatic operation.(BA)

Firmly mounted and yet mobile; that was the 128mm Flak on a special railway flatcar. In this way it could be moved quickly, especially to supply or freight yards in large cities. As with the 105mm Flak, the sides of the 128mm Flak car could be folded out. The ammunition was stored at both ends of the car. When no combat action was expected, the guns were set at 85 degrees to take weight off the spring equalizers. (2 x BA)

A "Flak tower", as this concrete bunker was popularly known, is being built here. A Flak battery with four guns will be permanently attached atop it. The lower part of the bunker will be a bombproof air-raid shelter for civilians.

Below: This 128mm twin Flak is mounted on a Flak tower. On the loading platform for the right barrel stand the loaders K13 and K12. The K15 beside them on a special platform sets the firing data for the fuze range received by telephone or, set here in a firing drill, with a handwheel. The seated gunner, K11, operates the traverse aiming machine.

Firmly mounted and yet mobile; that was the 128mm Flak on a special railway flatcar. In this way it could be moved quickly, especially to supply or freight yards in large cities. As with the 105mm Flak, the sides of the 128mm Flak car could be folded out. The ammunition was stored at both ends of the car. When no combat action was expected, the guns were set at 85 degrees to take weight off the spring equalizers. (2 x BA)

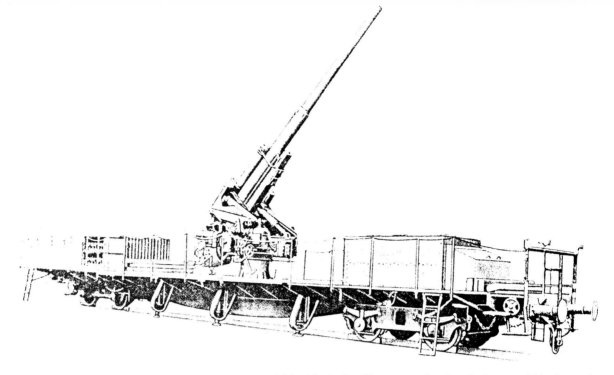

Above: The low-loader car for the 128mm Flak was stabilized by hydraulic rams under the platform and blocks on the wheels when it was in position.

Below: A 128mm railway Flak battery on the Channel coast in France. With the shells weighing 47.7 kilograms each, the ammunition loaders had heavy work to do when in action.(BA)

The 128 twinned Flak 40 was used chiefly on Flak towers to protect locations in big cities.(BA)

A "Flak tower", as this concrete bunker was popularly known, is being built here. A Flak battery with four guns will be permanently attached atop it. The lower part of the bunker will be a bombproof air-raid shelter for civilians.

Below: This 128mm twin Flak is mounted on a Flak tower. On the loading platform for the right barrel stand the loaders K13 and K12. The K15 beside them on a special platform sets the firing data for the fuze range received by telephone or, set here in a firing drill, with a handwheel. The seated gunner, K11, operates the traverse aiming machine.

The picture above shows the front of a twin Flak gun. In the switch box are the main and firing switches, the switches for elevation, traverse, fuze setter and rammer, as well as for both fuze power receivers. The K21 was responsible for them. The four equalizers and the two toothed arcs are striking.(BA)

Below: The guns could not be fired at this barrel elevation. As left sits the traverse gunner, the K11, at right the elevation gunner, the K1, and the K21 stands in front at the switch box.(BA)

Thick armored doors protect the way to the gun positions on the Flak towers.(BA)

This 128mm Flak twin stood on the Flak tower at the Berlin Zoo.(BA)

Here the rotating mount of a gun in a fixed position and the two sets of loading gear, each with a loading platform, can be seen clearly.

A 128mm Flak twin at high barrel elevation, with both loading platforms swung out. An independently functioning fuze setting machine with a fuze placing tray and an electrically functioning loading tray was mounted on each side.

At the lower left the crewmen's places when firing by electric transmission from the command device are shown: K1 elevation aiming gear, K11 traverse aiming gear, K2, K3, K12 and K13 loaders, K4 and K14 to handle empty cartridges with asbestos gloves, K21 to operate the switches on the switch box, K5 and K15 to set the fuze range according to telephonically transmitted data. The rest of the crew were ammunition movers; GF was the section leader.

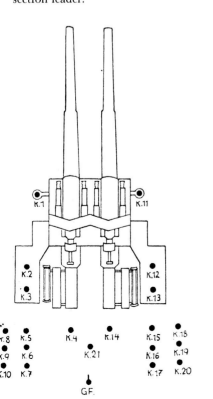

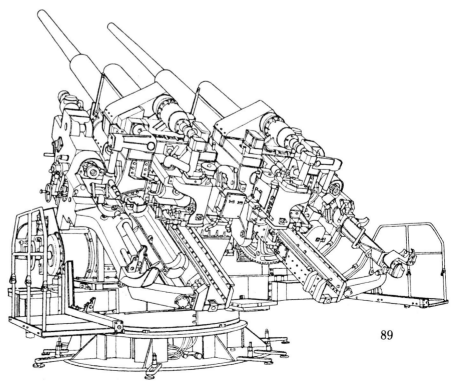

At a barrel elevation of 50 to 88 degrees, the loading platforms were swung out. From an elevation of 65 degrees on, the shell had to be pushed from the placing to the loading tray to prevent it from falling out. The size of a shell can be seen clearly in the picture.

Below: Here the placing tray with the fuze setting head, which moved to the fuze after a shell was placed on it, is easy to see.

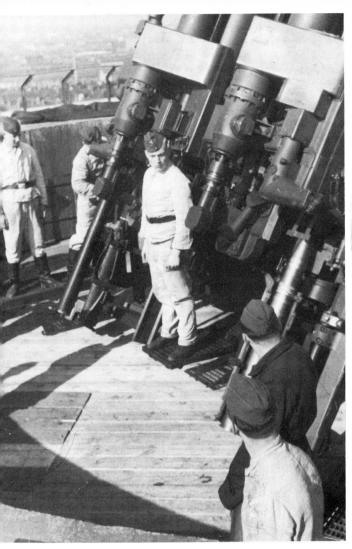

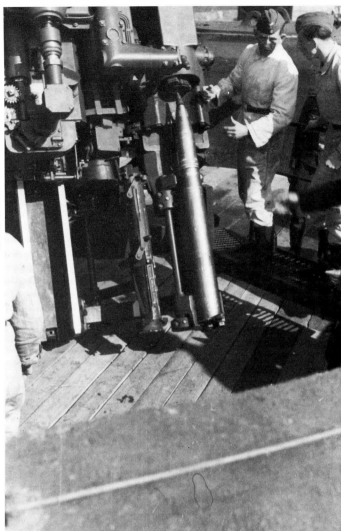

Upper left: A practice shell has been placed on the placing tray and the fuze setting head moves onto the fuze.(BA)

Upper right: After the fuze setting procedure, the shells are moved to the loading tray automatically. They swing in and the rollers of the rammer move the shells into the barrels.(BA)

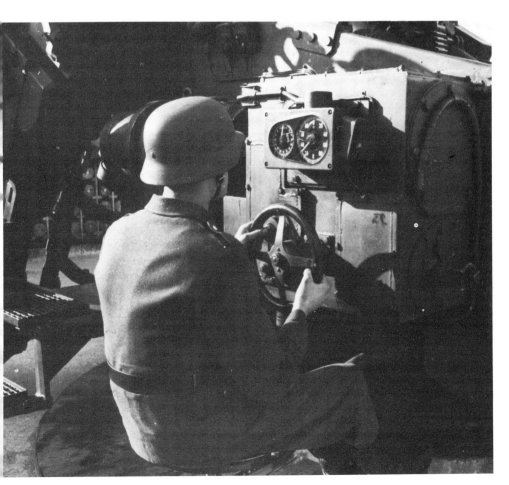

Left: By controlling the aiming speed, the traverse gunner matches the command indicator to the direction indicator.(BA)

Above: This 128mm Flak twin stood on a Flak tower in Berlin in 1943. Here again, the right gun can be seen very clearly. At the far right in the picture is the traverse aiming gear with the direction indicator receiver. Whereas the single 128mm Flak gun could be aimed either by hand or by motor power, the twin guns could be aimed only by motor. Thus there was only one handwheel present, for controlling the aiming speed. On the fuze setting machine (center) can be seen, among others, the switches for the fuze power receiver and the handwheel for setting the fuze range according to telephone transmission of firing data from the command device. (BA)

Left: During loading drills with practice shells, as in this picture, the barrel elevation could not be higher than 50 degrees. Unloading and opening of the breechblock took place with the help of a line that an ammunition loader had to operate.

Right: The gun crew of a 128mm Flak twin during loading drill.(BA)

Below: The section leader is in touch with the command post through his headset. On the left lower arm of his field jacket he wears the anti-aircraft emblem, which was awarded to non-commissioned officers and enlisted men for good achievements in combat.(BA)

On a Flak tower in Berlin.(BA)

An aircraft spotter on the Flak tower of a 128mm twin Flak battery.(BA)

As the 88mm Flak in slightly modified form was installed in Jagdpanzer tanks as a pure antitank gun on account of its good qualities as an armor-piercing weapon, so was the 128mm Flak. It was installed in the Jagdpanzer VI, the "Jagdtiger", as the 128mm Pak 44 L/55. Here are two pictures of the "Jagdtiger" with the 128mm antitank gun. This 128mm antitank gun could carry 38 shells on board. It was the best German antitank gun, and no enemy tank could withstand it.

150mm FLAK

Very few could have known that as early as 1938 prototypes of a 150mm Flak had been built by both Krupp and Rheinmetall.

The 150mm Flak 50 gun made by Krupp stood on a socket mount with four extenders. It had electric-hydraulic aiming machines, a fully automatic loading system, a roller rammer and an ammunition lift. The rate of fire was stated at 9 to 10 rounds per minute. The gun was portable when dismantled into four pieces (platform, lower and upper mounts and barrel). The later type was portable in three sections (platform, mount and barrel).

As for the 150mm Flak's ballistic data, the following is known:

Maximum range: 21,000 meters
Maximum altitude: 15,200 meters (only 400 meters higher than the 128mm Flak)
Muzzle velocity (Vo): 890 meters per second
Shell weight: 40 kilograms

Because of the great weight of 22,200 kilograms in firing position and comparatively high material and production costs, only a slight increase in performance above that of the 128mm Flak was possible, and thus the 150mm gun was not put into series production.

An improved 150mm Flak 60F by Krupp and a 150mm Flak 65 by Rheinmetall were not put into production either. For these guns the shot weight had been increased to 42 kilograms and the muzzle velocity to 960 and 1200 meters per second respectively.

A test model of the 150mm Flak.

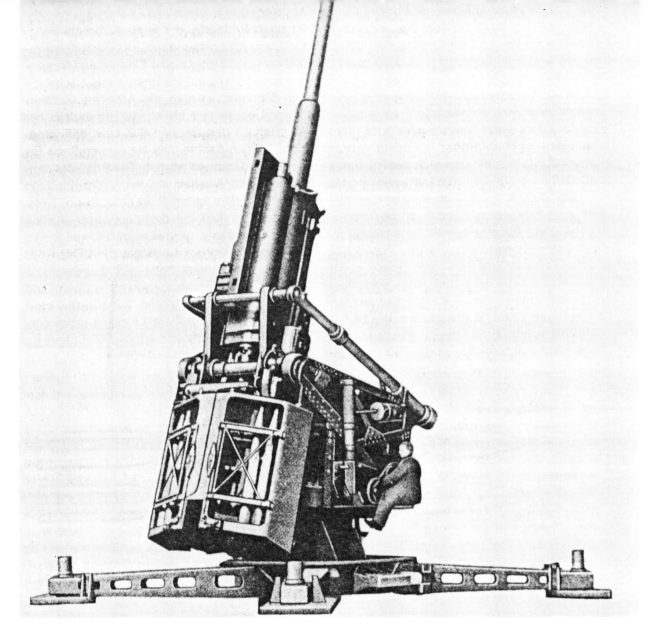

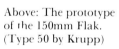

Above: The prototype of the 150mm Flak. (Type 50 by Krupp)

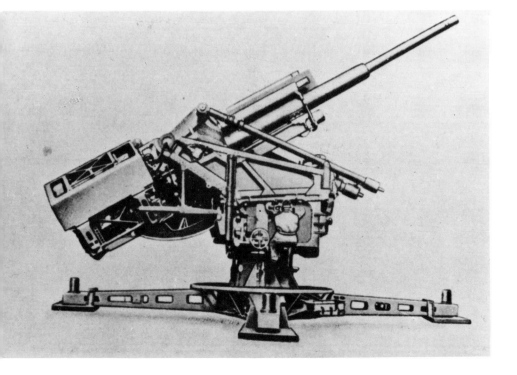

Left: The prototype of the 150mm Flak. (Type 60F by Krupp)

COMMAND DEVICES (Kdo.Ger.)

Kdo.Hi.Ger.   Kdo.Ger. 36   Kdo.Ger. 40

Unlike the Field Artillery, the main task of the anti-aircraft troops consisted of fighting fast-moving targets in three-dimensional space. In the process, the target speed, course and altitude changes had to be taken into consideration, in addition to influences on the shell trajectory such as wind direction and velocity, Vo (= muzzle velocity of the shell), powder type and temperature, shell weight class and much more, in order to calculate in advance, and as accurately as possible, the point at which the target and the adjustable-fuse shrapnel shell would meet. In order to calculate that, complicated measuring and calculating devices were necessary.

In World War I, from 1916 on, the distances were measured with 2- or 4-meter inverted-image range finders and the shells were fired according to data estimated with the help of range conversion tables. So-called lead computers then became the first fire-control devices. For more exact measurement of distances, experiments were made with 1-, 2- and 4-meter stereoscopic range finders. For this the man using the device had to have a stereoscopic, three-dimensional view.

After the development of fire-control devices for all the German Army's anti-aircraft weapons was halted by the Treaty of Versailles, a command device was developed in 1927 under contract from the Navy: the Kdo.Ger. "Pschorr 27." But only a small series was made available to the troops.

The KOMMANDO-HILFS-GERAT 35 that was developed afterward operated on angle velocities with the help of friction drives and graphs. The calculator and 4-meter range finder were positioned separately in the center of the gun battery. The firing data thus attained for traverse, elevation and range were transmitted to the guns by telephone. This resulted in inaccuracies to a greater or lesser degree, depending on the gun crew's level of training. But the Kommando-Hilfs-Gerät, the Befehlsstelle II (BII for short), could be achieved only after the development of the KOMMANDOGERAT 36, a further development of the Pschorr-Gerät 27, which was supplied to the troops in 1936.

The Kommandogerät 36 worked on a geometrical-linear basis. The 4-meter range finder (Raumbild-Höhenmesser = Stereoscopic Height Gauge, abbreviated Em 4m R(H)), was mounted

on top of the calculator. The firing data were transmitted electrically by a 108-strand remote-control cable to a distributor box in the center of the battery and on to the lamp receivers of the guns. If the lamp indicator system were put out of commission, telephone transmission to the guns was possible. The command device was located 500 to 600 meters away from the center of the battery, so as to enable its crew to work without disturbance. The difference in location, as well as the influences mentioned above, were included in the calculations when the device transmitted its firing data. Despite the precise operation of this device, errors caused by the 13-man crew needed to operate the device could not be avoided. In order to eliminate this source of error as much as possible and allow more accurate fire on targets with curving courses and changing altitudes, the Zeiss firm was occupied since 1937 with the development of a device that was intended to replace personnel with mechanical components including more than twenty electric motors, switch boxes, graphs, differentials, friction drives and regulators. In 1940 a command device, originally called KAPPA-GERAT, reached the troops; it required a three-man crew on the stereoscopic range finder 4mE(R) 40, two men on the calculator connected to it, and one man at the collector with the controls and apparatus for operating and transmitting power. The name Kappa-Gerät was given to the new command device from the Greek letter kappa, which indicated the angle of course, which this device was able to take into consideration, in the theory of anti-aircraft fire. The later designation of this device was Kdo.-Ger. 40; later improvements were Kdo.-Ger. 40B and 40C.

The Kdo.-Ger. 41 was similar to Kdo.-Ger. 40 but was based on the ballistics of the 88mm Flak 41. The Kommando-Hilfsgerät 35 was portable on a single-axle Special Trailer 35, the Kommandogerät 36 on the two-axle Special Trailer 104, and the Kommandogerät 40 on Special Trailer 52. The 4-meter base plus all equipment such as cables, collectors, containers for aiming telescopes and such, could fit into Messtrupp-Kfz. 74 or the 8-ton towing tractor (Zgkw) Kfz. 7/6, which had a special rail in the middle of the cargo surface, onto which the 4-meter base could be loaded in its transport case.

Here data from a lead computer are transmitted to the guns by telephone.

With this 2-meter base the range was determined in the early Thirties.

The Kommando-Hilfs-Gerät: a) Aiming telescope, b) Aiming handwheel, d) movable carrying bars, c) screen. Ready for transport, the device weighed 905 kilograms, ready for action 195 kg.

In the Kommando-Hilfs-Gerät was the 4-meter stereoscopic range finder was separated from the calculator. The measured distances to targets were transmitted by telephone and entered there.(BA)

The Kommando-Hilfs-Gerät was placed in the center of a Flak battery. With the 4-meter stereoscopic range finder, the 4-meter base (right foreground), the distance was measured. The B4, at the right edge of the picture, sent the distance to the calculator by telephone, in the left background. Since the target was also sighted optically by the calculator, it could calculate the shot traverse, elevation and fuze range, which were transmitted to the guns by telephone. The Kdo.-Hi.-Ger. 35 was independent of electric devices other than a battery for telephonic transmission, but remained only an auxiliary device. What with a 13-man crew and telephonic transmission of the firing data to the guns, too many human errors were possible.

A stereoscopic range finder 4 m Em (R) 40 used by an Army unit in Norway during 1943. At left is the E2. He aims the traverse at the target. In the middle is the E-Measurer, the E1, and at right the E3 is aiming at the target in terms of elevation. The gunners are also looking through a double telescope.(BA)

Above: A Messtrupp II rides to work on a ranging team truck, Kfz. 74, with a Kommando-Hilfs-Gerät 35 on a Special Trailer 53. The stereoscopic range finder, 4 m Em (R) H, or the 4-meter base for short, can be seen in a box in the center of the truck. It was loaded onto the truck by rolling the box along rails in the rear bed. On the back of the cab a cable drum is attached to hold the remote-control cable.(BA)

Right: "Air raid!" The crew rushes out of their bunker to the range finder. We see part of the Em 4 m (R) H on its fixed base.(BA)

The Kommandogerät 36 was developed from the Kommandogerät Pschorr 27. The 4-meter range finder mounted on it provided the device with the target height or distance. In the picture is the front of the device, with the drum that gave the barrel elevation at right, the fuze range drum at left and the adjusting knobs for making improvements in the middle.

On the right side of the device we see the handwheel (3) for elevation aiming, the bracket (L) for the 4-meter range finder, the transmitter arm (T & 45), the settings for altitude and vertical speed (h), the handwheel (6) for matching the target altitude curves (7 & 5) and the removable braces (P) for the protective cover.

The picture above shows the back of the Kommandogerät 36. At right is the handwheel for the elevation aiming gear (3), the aiming telescopes (S) for coarse and (F) fine aiming at the target, the scope (E) for the range measurer with a handwheel (M) for range measurement, the handwheel (2) for the traverse aiming gear with the telescopes (S) and (F) for horizontal aiming.

The Kommandogerät 36 standing at the left on a firing range. (BA)

The three crewmen at the 4-meter range finder of Kommandogerät 36 aiming at a target. The E1 is not wearing a steel helmet; beside him is the E2 at the traverse aiming gear, and to the right of E1 is E3, aiming at the height of the target. The crewman in the foreground is the B6. He stands at the flight direction table on which the indicated movements of the plane must be correlated.(BA)

For the "Technical Shooter", whose headset and microphone keep him in touch with the guns, there were binoculars on the 4-meter range finder.

Here again, the B6 is at work at the flight direction table, while the "Shooter" observes the air target through his binoculars.(BA)

Advancing through the snow was difficult with the Kommandogerät in tow. The Kdo.Ger. 36 is pulled by the range team truck Kfz. 7/6, an 8-ton towing tractor. It had rails in the middle of the rear bed on which the 4-meter range finder was loaded. In the boxes at the rear of the tractor are the storage battery, the loading gear, cable drums and other accessories.

Here the transport case is open, showing an improved 4-meter stereoscopic range finder, the 4 m Em (R) 40. With the help of the rollers on the bottom of the case, the range finder could be loaded onto the rails of the towing tractor more easily.

A collector with a switch box that provided the operating power for the Kdo.Ger. 36 and also the Kdo.Ger. 40, plus the power for the Transmission Device 30 or 37. The tension was between 34 and 45 volts.

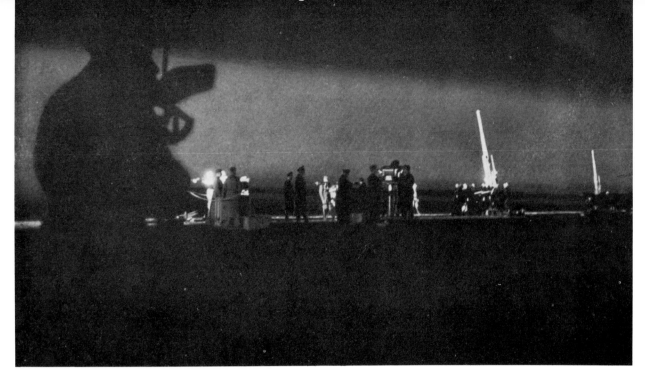

Above: Night firing by an 88mm Flak battery with a Kommandogerät on a firing range before the war.

Below: A camouflaged Kommandogerät 36 on a drill before the war. This type of camouflage was no longer used during the war, since it took too long to remove in case of an air raid.

The Kommandogerät 40 with the stereoscopic range finder 4 m Em (R) 40 was first delivered to the troops in 1940. The 4-meter base had a measuring range of 1200 to 100,000 meters. The 20-power optical enlargement could be switched to 32-power. At the left i the back of the device. The measurer stood on a platform in the middle and was turned with th device while using it. The traverse gunner stood to the right of the E the elevation gunner to his left. The binoculars under the left side of the base were for the "Technica Shooter." At the left rear side of th device is the securing box of the 2 electric motors that worked in the device. The crew consisted of a range measurer, two aimers, two crewmen on the device and one man at the switch box. The upper left device is there for optical testing, for theoretical firing in which the data are recorded and later evaluated.

The middle picture shows the Kd Ger. 40 in position. The bed is expanded by a wooden grid, so tha the shooter's binoculars could be mounted on top of the range finder. The crewman in the pictu stands at the position of the B4, who used a handwheel to aim by correlating curves. The many dial allow various data to be read or photographed for optical testing.

The bottom picture shows a covered Kdo.Ger. 40 in position. Next to it is the case for the 4-met range finder.

Generalleutnant Schmundt and Oberst von Natzmer, a bearer of the Knight's Cross and First General Staff Officer of the Panzer-Grenadier Division *Grossdeutschland*, inspecting a Flak battery in the east in 1944. The picture also affords a look at the top of a Kommandogerät 40. Under the rectangular glass panel was the hand course drum with the course recorder, over it the aircraft viewing screen and the course indicator. Under the housing of the panoramic telescope are the indicators for horizontal speed, the Vh, the round main switch in front of it, and the switches for course and altitude changes.

Below: Only the E1, the two aimers and the "Technical Shooter" are standing at this device; the Shooter holds the cable with the fire bell. It gave the guns the command to fire.

Above: Here the traverse aimer, the E2, stands at his binoculars with the coarse viewer above them. At left under the binoculars in the indicator window for the measured distance. From the painted number 4 a power cable runs to the aiming telescope and range-finder scale for night illumination.(BA)

Below: The measurer is the man without the steel helmet, which would disturb him at his work. At left is the aiming telescope 3 of the elevation aimer, in front of it an indicator for fuze range, then the B4 and B5 are at the main switch. From the frame over the device, one power cable comes from the range finder and one apparently leads to the "Malsi" evaluating device.(BA)

The Kdo.Ger. 40 at left is standing on the Flak tower of a 128mm Flak battery in Berlin.(BA)

Lower left: The Kdo.Ger. 40 is carried on the single-axle Special Trailer 52. The side sections of the bed are folded up.

Below: On the back of this Kdo.Ger. 40 is a box with the receiver part of the Transmission Device 37 with direction indicator. Here the data for distance, traverse and elevation provided by the range finder were received. They had to be correlated by the aimers, more or less as at the guns.

A very interesting gathering of heavy Flak guns and devices on a firing range. In the right foreground are four Kdo.Ger. 40, two with light and two with dark camouflage. To their left are the boxes for the 4-meter range finders and two covered battery banks each. Two panoramic telescopes to connect the Kdo.Ger. with the guns stand on standards. Farther left are three motor generators to power the collector sets. On the rails are, at left, a special car with a radar device for the 128mm railroad Flak, the guns of which are in the background. The guns of three 88mm Flak batteries are hard to see, partially because of their bright camouflage paint.(BA)

Below: The Kdo.Ger. crew of the railway Flak unit working together on the range finder. Here the data are still received by the aimers by telephone and put into the Kdo.Ger.(BA)

RINGTRICHTER-RICHTUNGSHORER
(RRH)

RINGTRICHTER-RICHTUNGSHORER (RRH), in short, listening devices, were issued to the Flak floodlight units in 1937. By amplifying the human ability to hear to as far as 12 kilometers, and by being connected to calculators, they were to allow faster aiming of floodlights at air targets. Firing according to sound with the help of RRH but without the use of floodlights was, for a time in 1940, preferred to ammunition-wasting barrage firing. The results, though, were unsatisfactory in view of the higher- and faster-flying enemy planes. The auditory range of 5 to 12 kilometers was dependent on the weather, the sound level of the plane to be located, and the listener's capability. But as long as there were no radar devices (FuMG), which even later were not available in sufficient numbers, one had to be satisfied with the RRH when firing in fog, under low cloud cover or at night without floodlights. That is why there were still 5559 listening devices in service up to 1944.

This trumpet shaped direction finder consisted of a 723-kilogram bedplate, a socket, a column, two arms, the trumpet, the traverse and elevation controls and the calculator. The RRH, including the calculator but without the bed, weighed 983.3 kilograms and was portable on the two-axle Special Trailer 104.

A listening troop consisted of a troop leader, the K 10, a man to run the computer, the K 7, who stood at the back of the device, a listener for the traverse trumpet, the K 8, on the left side of the device, and a listener for the elevation trumpet, the K 9, on the right side.

RRH, bed offset

The right arm of an RRH

Components

3 Bedplate
7 Hook for lifting chain
8 Opening for centering pin
9 Trunnion
10 Leveling jack
11 Ring
12 Joint bolt
189 Toolbox
190 Equipment box

Components:

52 Attachments
53 Rubber buffer
54 Pillow block
55 Rubber middle block
56 Eyes
91 Acoustic elevation loop
92 Acoustic elevation indicator
93 Pushrod
178 Elevation indicator lamp
181 Elevation signal box
191 Rain cover ring
192 Crossbrace
193 Clamp

111

The front of an RRH.

13 Nameplate
41 Lid for small opening
45 Mounting rods
46 Mounting rod brackets
48 Ball handle
49 Seats
50 Footrests
51 Handholds
77 Elevation handwheel
78 Traverse handwheel
149 Traverse angle indicator

Back view of the RRH column.

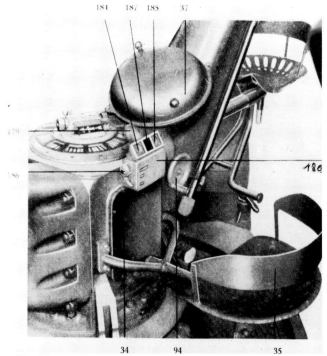

34 Lid
35 Calculator man's seat
37 Protective cover
94 Calculator control
179 Calculator lamp
180 Calculator signal box
184 ''Elevation hears'' signal
185 ''Traverse hears'' signal
186 Push button
187 Non-glare disc

The rear of an RHH with seats for listeners under a rain cover. The calculator is not aimed.

Components:

5 Bed Hook
6 Locking jack
14 Socket
23 Column
42 Right arm
43 Left arm
57 Locking Rod
58 Locking rod spring
59 Trumpet
150 Elevation angle indicator

Drawing showing dimensions of RRH

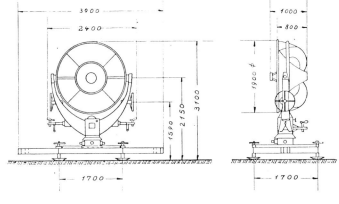

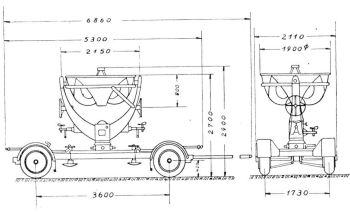

This RRH has been placed along a road. The device has been lowered from the Special Trailer 104. The two listeners for elevation, the K9, and traverse, the K8, have put on their headsets and are trying to locate the target. (2 x BA)

Above: This listening device was set up by the coast guard south of Boulogne. The traverse listener, the K8, sits at left, the listener for elevation, the K9, at right. In the middle, the K7 stands at the calculator.(BA)

Below: Listening devices used in drill before the war.

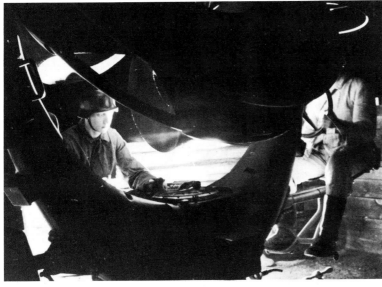

Above: Foreign officers are shown how a listening device works.(BA)

Center: Listeners on night duty. The K7 sits at the calculator doing his work.(BA)

Right: This listening device, carried on a Special Trailer 104, is being taken to its position.

Above: This RRH has been positioned to support a floodlight, at left behind the RRH. The rain covers are around the seats of the two listeners.(BA)

Below: Floodlight and RRH working together in tracking air targets.(BA)

The elevation listener sits before the acoustic elevation arc and watches the acoustic indicator while controlling the device with the elevation handwheel.

FLOODLIGHTS 60-cm, 150-cm, 200-cm

In World War I, from 1917 on, floodlight crews were trained at a floodlight school. 60-cm, 110-cm and 200-cm floodlights were already in use. But here too, after the war the Treaty of Versailles required a cessation of development until 1927. After that, in the Thirties, the 60-cm FLOODLIGHT with an 8-kilowatt generator was developed, especially for light and medium Flak, and was improved in World War II.

The crew consisted of a floodlight leader, an aimer, the K 1, in the aiming seat, the lamp tender, the K 2, who operated the lamp controls for bundling and scattering light and was responsible for the fuel supply, and the machinist, the K 3, who operated the motor and switched the power to the machinery on and off. The floodlight was portable on a single-axle Special Trailer 51.

Instead of the 110-cm floodlight, the 150-cm Flak Floodlight 34 and 37 were used; they had self-regulating high-performance lights and a power of 1.1 billion Hefner candlepower, a parabolic glass reflector, the Darkness Searching Device 41, and a 24-kilowatt generator. (One Hefner candlepower is the unit of light produced by one normalized amyl-acetate wick lamp.)

This floodlight had a crew composed of a leader, a lamp operator, the K 2, the two aimers, K 1 and K 5, the lamp tender and the machinist. Both the floodlight and the 24-kilowatt generator could be carried on two-axle Special Trailer 104 units.

In 1943 the first 200-cm SCHEINWERFER 43, with a power of 2.7 billion Hefner candlepower, a 120-kilowatt generator and Darkness Searching Device 41 was supplied to the troops. With it, targets up to a distance of 12 to 13 kilometers could be spotted.

The use of floodlights in fog, mist, cloudy weather, and even on moonlit nights was often unsatisfactory or downright impossible. But before the introduction of radar devices (FuMG) they were indispensable for the optical spotting of air targets at night. Even after radar was introduced, they could not be dispensed with, as the enemy often nullified the radar devices by causing interference. Among other uses, floodlights were used along with radar, which was used to aim them at targets. The elevation and traverse aiming data were transmitted by Transmitting Device 37.

The supply of floodlights in February of 1944, according to statistics supplied by the Quartermaster General of the Luftwaffe General Staff, was:

60-cm Flak floodlight-5582 mobile, 794 fixed,	total 6376
150-cm Flak floodlight-5675 mobile, 1636 fixed,	total 7311
150-cm quadruple Flak floodlight-61 fixed,	total 61
200-cm Flak floodlight	total 2262

The picture shows a 60-cm floodlight in position. The lamp operator, the K 1, sits in his aiming seat; the lamp tender, the K2, stands behind him.

A 60-cm floodlight in position in the field. The K1, in his seat, aims the floodlight according to the instructions of the leader, who observes the target through his binoculars. The K2 stands behind the floodlight. He is responsible for supplying fuel and operates the lamp controls for bundled and scattered light.

On this 60-cm floodlight the elevation and traverse aiming cranks before the seat are easy to see.(BA)

150-cm floodlights and trumpet listening devices at the weapons show in Berlin on Hitler's 50th birthday in 1939.

A 150-cm floodlight in position. At right is the elevation handwheel and left the projecting traverse handwheel with the chest tiller. The shield of this floodlight is still closed.(BA)

Here we see the left side of a 150-cm floodlight on the Channel coast, with the elevation handwheel, elevation arc and cables from the generator.(BA)

A 150-cm floodlight is put in position. One limber of the Special Trailer 104 has been detached and the traverse controls folded out.

Here a floodlight position is being prepared.(BA)

A floodlight crew working at night.

Japanese officers
inspect a
floodlight
battery.(BA)

The shield of this
200-cm floodlight
was opened for
inspection of the
big parabolic
mirror.(BA)

The generator for a
floodlight is being
shown.(BA)

Floodlights in use. Of necessity, their positions were lit as brightly as in daylight and thus offered easy targets for low-flying planes and bombers.(BA)

Before radar came into use, spotting and engaging aircraft at night was possible only by using floodlights. Unfortunately it proved that the cooperation of trumpet listening devices, floodlights and gun batteries was not as successful as had been thought in peacetime. In addition, the floodlights set up around a protected location often provided the enemy with a target when they were lit. Nevertheless, the use of floodlights could not be dispensed with during the war.

Blazing floodlights and exploding anti-aircraft shells often lit up the night sky over German cities during the war.

Bombers flying at ever-higher altitudes demanded
ever-stronger floodlights. Thus the 200-cm floodlight
was put into service in 1943. With it, targets up to a
distance of 12 to 13 kilometers could be spotted. Here
are three pictures of this 200-cm floodlight. (2 x BA)

126

During the course of the war, more and more well-trained soldiers were transferred from the home batteries to fill vacancies in front-line units. They were often replaced by schoolboys and Hitler Youth members as Luftwaffe helpers or Flak helpers. In intelligence and floodlight units or ranging units of gun batteries, female Flak helpers and Arbeitsdienst (Work Service) girls were also used. Above: Women Flak helpers at a 150-cm floodlight.(BA) Lower left: Arbeitsdienst girls at a floodlight. Lower right: Flak helpers at a 200-cm floodlight.

RADAR DEVICES (FUMG)

Successful action against air targets was possible only with exact measurement of the target's range, vertical and horizontal location. With the help of the stereoscopic range finder at the command device, one could determine this only under certain conditions: the target had to optically visible and not obscured by fog, smoke, clouds or darkness, and the spatial sight and training level of the spotter had to be good.

In 1940 the first radar devices, FuMG for short, reached the troops. They were independent of optics. The calculation of the target range was accomplished by an echo-sounding operation, in which electronic impulses were broadcast, were reflected when they hit a solid body and became visible on a scope as fluorescent lines or spots. The range data thus obtained were considerably more exact than those optically obtained. Unfortunately, the development of these devices in Germany was at first very slow, so that the advantage gained by the English in the realm of radar technology could scarcely be attained by the end of the war.

By 1945 fifteen different radar devices had been developed and used in Germany, of which only the best-known can be listed briefly here:

FuMG Freyz (A1) was a panoramic searching device and was used mainly as an early warning device.
FuMG 39L, 40L (A2) were navigational devices made by the Lorenz firm that could also be used for gunfire.
FuMG 62 (39T "Würzburg" A-D), of which some 4000 were made.
FuMG 63 (40T); in this "Mainz" device, as in
FuMG 64 (41T), this "Mannheim" device the data were transmitted automatically to the command device by Transmitting Device 37.
FuMG 65 was set up and used in large batteries, on Flak towers and in mobile form on railroad cars under the name "Würzburg Giant."
FuMG 64 "Ansbach" and FuMG 74 "Kulmbach" were used only in small numbers, as were the FuMG 75 "Mannheim Giant", FuMG 76 "Marbach" and FuMG 77 "Marbach V."

This FuMG is surrounded by a screen. Behind it is a covered floodlight.(BA)

The most frequently used device, the FuMG 39T "WURZBURG", had a two-section antenna dish of three-meter diameter. In the center was a circulating dipole on a conical column, equipped with a protective cover. The dish was mounted on a turning turret on a socket. On the turret itself were the sending, receiving and net cabinet and two seats for the elevation and traverse observers. On the right side of the turret was the so-called extender with a seat for the B2, equipped with leg protectors and foot-warmers as well as foot pedals and a traverse aiming handwheel. The viewing tube with range calibration, the traverse and elevation aiming tubes were also part of the extender.

The precision of measurement from +/-80 to +/-120 meters was at first insufficient for providing firing data. It was soon improved to +/-25 to +/-40 by the D-Zusatz auxiliary equipment. The vertical and horizontal aiming accuracy were also improved more and more, so that the device finally provided very useful data

to the command device. The data for elevation, traverse and range, originally transmitted by telephone, were later transmitted to the command device electrically by the Direction Indicator System 37. The weight of the device amounted to 1500 kilograms. It was portable on Special Trailer 104, with the two single-axle limbers interchangeable.

The larger FuMG 41T "MANNHEIM" was portable on Special Trailer 204, likewise with the two single-axle limbers interchangeable. The device stood on a cross mount. In its closed operating condition there were benches for B1, B2 and B3 complete with foot-warmers. The rear wall of the cabin consisted of a tarpaulin; the sidewalls had doors with windows.

Because the supply of these two devices was far below their demand, as of August 1944 several large batteries were linked to a single device.

After a British commando operation had seized the most important components of a German FuMG in 1942, The British developed means of interference to put these devices out of operation. In a large-scale attack on Hamburg in 1943, lasting for several days, great quantities of aluminum foil, called "Düppel" in Germany, were first thrown from aircraft; they appeared on the radar scopes as lines of light, thus making the identification of air targets impossible. By using accessory devices such as the "Würzlaus", "Tastlaus", "Taunus" and "Windlaus" or "K-Laus", attempts were made to nullify this interference, which practically resulted in a war of radar technicians.

No more can be said within the framework of this book on the development of radar devices without making this chapter too long.

This picture shows an FuMG 39T (D2) on a wooden platform extending some 35 to 45 centimeters around the device. An alarm has sounded, and the crew hurries to their places. At left, the B6 jumps into his seat before the elevation arc to transmit the elevation. The B5 hurries to the horizontal angle indicator. The B1 stands at the fine range measuring device in front of the control box. The B3 stands in front of the vertical aiming tube and the B2 sits in the aiming seat at the far right.(BA)

This FuMG 39 T is part of an 88mm Flak battery in service on the northern coast of France. The position, built up with sandbags, offered the crew very little shrapnel protection as they worked. The two-section dish had to stand above the breastworks, as clear measurement was not possible otherwise. At right the cable runs to the fire-control position, Command Post 1. Behind the FuMG are the gun emplacements, so laid out that all four guns had a free field along the coast and the ocean.(BA)

Here is an FuMG 39 T position in the mountains of Feodosia.

At right is an FuMG 39 T (D2) on a wooden platform. The two-section dish on the rotating mount had a diameter of 3 meters. In the middle is the rotating dipole on the conical dipole pillar, with a protective cap. The B2 sits at the extender in his steering seat with leg protectors, into which foot pedals and foot-warmers have been built. In front of him is the traverse aiming handwheel, and the traverse aiming scope is at his eye level. His headphones were used along with the "Steinhäger" or "Nürnberg" auxiliary device which was used to counteract enemy jamming.(BA)

Lower left: The B2 observes the horizontal light lines in his traverse aiming scope, while the B3 beside him monitors the vertical light lines in the elevation aiming scope.(BA)

The dish of this FuMG has painted on it the score of planes its crew has helped to shoot down.

This FuMG is part of the 39 T (D4) "Würzburg" version. These devices had a metal grid 65 to 75 centimeters above ground instead of a wooden platform. The sensitive dipole has a small protective cover on it.(BA)

At left, the B2 again sits in his aiming seat before the horizontal aiming scope. Behind him is the troop leader with headset and microphone. Beside him is the B3 at the vertical aiming scope.(BA)

Below is a drawing of the control box.

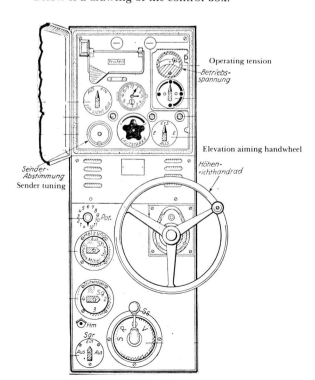

Operating tension
Betriebs-spannung

Sender-Abstimmung
Sender tuning

Elevation aiming handwheel
Höhen-richthandrad

Above on the special Flak railroad car an FuMG 62 (39 T-D4) is mounted. The raised platform of the device extends well over the sides of the car when unfolded, so that the car could only be moved when there was no traffic on adjoining tracks. The crew's cabin is on the same car, near the device. The railroad Flak emblem can be seen on the wall of the cabin.(BA)

At right is a picture of the FuMG 62 "Würzburg" with its platform, rotating turret, elevation arc and seat for the B6.

For the air intelligence troops, the fighter command, as well as the 128mm Flak twins on the Flak towers, the FuMG 65 "Würzburg Giant" was very useful. Some of them stood along the Channel coast, so as to be able to sense and report enemy flights as early as possible. The picture above shows one of these devices near Boulogne before the invasion.(BA) Because of its great weight of eight tons, it could only be mounted permanently or on a railroad car. The diameter of the dish was 7.5 meters. The horizontal range, depending on the location, was 50 to 80 kilometers wide; the accuracy was +/-1.5 to 1.8/16 degrees.

Left: An FuMG 65 in operation.

The picture above shows the interesting "Egerland" FuMG setup. It consisted of a panoramic searching device, the FuMG 74 "Kulmbach", at right, which constantly sought the target area. When a target was found, the FuMG 76 "Marbach" at left was aimed at this target. The "Marbach" device then supplied more precise targeting data to the command device of a Flak battery. This FuMG setup existed only in small numbers.(BA)

Right: The FuMG 76 "Marbach" made by Telefunken. Its dish had a diameter of 4.5 meters. Its horizontal range was about 50 kilometers; its accuracy was +/-1/16 degrees both horizontally and vertically.

The "Freya" (A1) radar device made by Lorenz was a panoramic searching device and actually suitable only for use as an early warning system, but because of a shortage of more precise equipment, it was used in 1940 in a 105mm Flak battery near Düsseldorf to direct a command device and a floodlight that stood beside it. This was soon moved, as it illuminated the whole position when lit. The aiming data transmitted telephonically from the "Freya" device were, of course, not very precise, but at least the barrage fire could be aimed more accurately. The device was at that time strictly top secret. Other than its crew, nobody was admitted to the closed-off position of the device.(BA)

When the war began, the English very soon made daylight attacks, though infrequently, as the losses were high. After that they flew only night raids on cities in western Germany. The Flak guns' chances of success against these raids were low, as they were limited to fighting optical targets. Since sighting targets at night with the use of floodlights was more difficult than had been assumed in peacetime, better results could be achieved only when radar was used. Later, when the Americans flew day attacks with their four-engine B-17 bombers, the skies over German cities often looked, at times of heavy Flak fire, as shown in the picture above. At night, though, the British flew as before, and the Flak exploded as seen in the middle picture. The effect of the Flak, as seen in the picture below, was taken seriously by the enemy. The 9th U.S. Air Fleet reported that 49.6% of aircraft losses and 92.9% of damage had

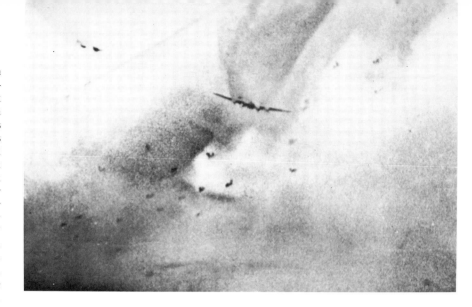

been caused by the German Flak. On the German side, the average ammunition consumption per month in 1944 was: For the 88mm Flak 1,749,300 rounds, for the 105mm Flak 199,800 rounds, for the 128mm Flak 73,700 rounds. Statistically, the shooting down of one four-engine bomber could be expected with the following average number of rounds:
16,000 rounds of 88mm Flak 36/37
8500 rounds of 88mm Flak 41
6000 rounds of 105mm Flak 39
3000 rounds of 128mm Flak 40

137

A typical day at a Flak position. The men stand on guard or at aircraft reporting posts but are ready to fire in seconds after the alarm sounds.

BALLISTIC DATA

GUN MODEL	88mm 18, 36, 37	88mm 41	88mm 37/41	105mm 38/39	128mm 40	128mm Twin Mounts	150mm 50/55
Caliber	88	88	88	105	128	128	150
Barrel length mm	4,930	6.548	7.027	6.648	7.835	7.835	—
Rifled section mm	4.124	5.411	5.850	5.531	6.478	6.478	—
Number of riflings	32	32	32	36	40	40	—
Barrel recoil mm	850 – 1.050	900 – 1.200	—	830 – 900	1.000 – 1.300	1.000 – 1.300	—
Fire control device	Kdo.Hi. 35 Kdo. 36 u. 40	Kdo. 40	Kdo. 40	Kdo. 36 u. 40	Kdo. 40	Kdo. 40	—
Transmission device	30 u. 37	37	37	30 u. 37	37	37	—
Arc of traverse	360°	360°	360°	360°	360°	360°	360°
Arc of elevation	-3° bis +85°	-3° bis +90°	-3° bis +85°	-3° bis +85°	-3° bis +87°	0° bis +87°	—
Rate of fire rounds/minute	15 bis 20	20 bis 25	15 bis 20	12 bis 15	10 bis 12	20 bis 24	9
Firing position length mm	7.620	9.658	7.700	8.420	15.000	9.120	—
Firing position width mm	2.305	2.400	2.400	2.450	2.500	5.045	—
Firing position height mm	2.418	2.360	2.600	2.900	3.550	2.950	—
Height of fire mm	1.600	1.250	1.600	1.800	2.300	—	—
2-axle special trailer	201/202	202	202	203	220	2-203	—
Weight for transit kg	7.200	11.200	9.300	14.600	27.000	—	—
Weight in position kg	5.000	8.000	7.111	10.240	18.000	27.000	22.200
V_0 explosive shell meters/sec	820	1.000	1.000	880	880	880	890
V_0 armor-piercing shell meters/s	795	980	980	860	860	860	850
Maximum shot distance meters	14.860	19.800	19.800	17.700	20.900	20.900	—
Maximum shot height meters	10.600	14.700	14.700	12.800	14.800	14.800	15.200
Fuze range meters	10.600	12.350	12.350	11.800	12.800	12.800	—
Explosive shell weight kg	14.7	19,2	19,2	26,0	47,7	47,7	—
Armor-piercing shell weight kg	15.3	19,8	19,8	26,1	46,5	46,5	—
Shell shot weight kg	9.5	9,4	9,4	15,1	26,0	26,0	40,0

Today one can see the 88mm Flak 18 (above) at the "Study Collection of Military Technology" in Koblenz. The 88mm Flak 37 (below) is at Army Anti-Aircraft Battalion 5 in Lorch on the Rhine.